国外计算机科学教材系列

双语版C++程序设计

（第3版）

Learn C++ through English and Chinese
Third Edition

［爱尔兰］ Paul Kelly 著

苏小红

电子工业出版社

Publishing House of Electronics Industry

北京·BEIJING

内容简介

本书由在计算机程序设计方面有着丰富教学和实践经验的中外作者合作编写，并在其第2版的基础上进行了修订与更新。本书内容共14章，由浅入深、全面介绍了C++程序设计方法。本书通俗易懂，所有实例经过精心挑选、贴近生活，尤其强调读者的亲自参与意识。大多数章都为初学者提供了常见错误分析，所选习题可提高读者上机编程的兴趣。

本书是国内首次出版的中英文对照混排式双语版C++程序设计教材的更新版，既方便初学者熟悉相关概念和内容，也便于英语非母语的读者熟悉英文专业词汇。

本书可作为高等学校计算机、软件工程和其他理工类专业的C++程序设计双语版教材，也可供程序员和编程爱好者参考使用。

版权贸易合同登记号　图字：01-2025-2198

图书在版编目（CIP）数据

双语版C++程序设计：英汉对照 / （爱尔兰）保罗·凯利 (Paul Kelly)，苏小红著. -- 3版. -- 北京：电子工业出版社，2025. 6. --（国外计算机科学教材系列）. -- ISBN 978-7-121-50295-8

Ⅰ. TP312.8

中国国家版本馆CIP数据核字第2025F0J259号

责任编辑：冯小贝
印　　刷：三河市良远印务有限公司
装　　订：三河市良远印务有限公司
出版发行：电子工业出版社
　　　　　北京市海淀区万寿路 173 信箱　邮编：100036
开　　本：787×1092　1/16　印张：22　字数：683 千字
版　　次：2010 年 6 月第 1 版
　　　　　2025 年 6 月第 3 版
印　　次：2025 年 6 月第 1 次印刷
定　　价：59.00 元

凡所购买电子工业出版社图书有缺损问题，请向购买书店调换。若书店售缺，请与本社发行部联系，联系及邮购电话：(010) 88254888，88258888。

质量投诉请发邮件至zlts@phei.com.cn，盗版侵权举报请发邮件至dbqq@phei.com.cn。

本书咨询联系方式：fengxiaobei@phei.com.cn。

Preface

This textbook teaches the fundamentals of programming using C++, a programming language which supports the development of software using the object-oriented paradigm.

Although the book is primarily intended as a textbook for a programming module in a computer science course, it is equally suited to an individual familiar with another programming language and who now wants to learn how to program in C++.

This book focuses on core concepts and features of C++ while keeping the explanations as simple as possible. Drawing on their professional experience, the authors teach C++ programming largely by way of examples that are organised for easy step-by-step learning.

Like so many other programming languages, C++ contains numerous English technical terms that are difficult for all students, including the native English speaking student.

Learn C++ through English and Chinese explains C++ concepts and terminology in English with additional explanatory annotations in Chinese. This bi-lingual approach will be appreciated by Chinese students and will help them focus on the C++ language without being over-burdened by English technical terminology. Despite C++ being available on a wide variety of platforms, this book is not specific to any particular machine, compiler, or operating system. All programs are designed to be portable with little or no modification to a wide variety of platforms.

Learn C++ through English and Chinese

- Is a comprehensive introduction to programming in C++.
- Uses practical examples to explain difficult theoretical examples.
- Uses a step-by-step approach with detailed explanation of programming examples.
- Uses explanatory annotations written in Chinese.
- Provides end-of-chapter "Programming pitfalls" commonly experienced by learners.
- Provides a "Quick syntax reference" at the end of each chapter that summarises the C++ syntax covered in the chapter. This is a useful resource for experienced programmers as well as for learners.
- Provides end-of-chapter exercises, allowing the learner to test and re-enforce their understanding of C++.
- Is suitable for students new to programming and those familiar with some other language, such as C or Basic, and who now wish to learn C++.
- Is accompanied by a web site containing the example programs, solutions to selected exercises, frequently asked questions and links to other useful resources.

Typographic Conventions

The line numbers to the left of the program examples are for reference purposes only and are not part of the C++ language.

When a new term is introduced it is in *italic* type.

`C++ statements, keywords, program variables and values are in this font.`

This font is used in examples to show values that should be typed at the keyboard by the user.

Paul Kelly
Technological University Dublin

前　　言

为了适应国内大学逐渐与国际接轨的发展趋势，英语教学和双语教学越来越受到人们的重视。一方面，在教育部的大力支持下，许多课程建设成为教育部双语教学示范课程，但是这些课程大多采用英文原版教材，而面向双语教学的双语版教材在国内实属罕见。另一方面，以"国际化、工业化"为办学理念，注重国际化、工业化人才培养的国家示范性软件学院的部分课程，还常常邀请一些外籍教师来国内进行全英语授课。但是，由于目前国内学生的英语水平参差不齐，导致全英语授课的教学效果并不是非常理想。本书正是在此背景和需求下应运而生的。

本书的第一作者是爱尔兰都柏林理工大学（TUD）的高级讲师Paul Kelly。Kelly长期从事程序设计类课程的教学工作，教学和实践经验丰富，在国外已先后出版多本程序设计语言类书籍。自哈尔滨工业大学国家示范性软件学院成立以来，Kelly多次作为外聘教师在该学院从事程序设计方面的教学工作，对中国学生比较了解，针对其在教学中发现的问题，即初学者面临着既不熟悉专业术语和基本概念，又不熟悉英文专业词汇的双重困难，提出了出版中英文对照混排式双语版教材的思路，帮助学生在克服语言障碍的同时，能够更快、更好地熟悉和掌握程序设计方面的基础知识，为国内的双语教学提供了一种很好的解决方案。

本书内容共14章，由浅入深、全面介绍了C++程序设计方法，既适合以C++作为入门语言的读者，也适合学习过其他程序设计语言想再学习C++的读者。本书的特点如下。

1. 使用非常实用和贴近生活的例子及图示来通俗易懂地讲解难以理解的概念，在介绍面向对象程序设计方法时，尤其强调读者的亲自参与意识。

2. 采用案例驱动和循序渐进的方式，从一个应用实例出发，先利用现有知识编写出一个较为简单的程序，然后在此基础上不断扩充，在扩充的过程中再引入一个新的概念和知识点，逐渐编写出一个较大的程序。对每个例程都有详细的讲解，有的章以一个例子为中心贯穿始终进行讲解，后面章节的例子还会重用已有的部分程序代码，前后章节之间既有内容上的联系，也有例程上的联系。

3. 对重点内容和段落给出了中文注解。

4. 大多数章都为初学者提供了常见错误分析，以帮助初学者在程序设计中避免这些错误。

5. 大多数章都有快速语法参考，总结本章知识点，便于读者快速查询相关内容。

6. 大多数章都有精心设计的、有趣的习题，便于读者测试和强化对相关内容的理解。

7. 有相关的教学网站（华信教育资源网，网址为http://www.hxedu.com.cn）和教材网站（参见华信教育资源网本书页面上的信息），方便读者下载示例的源代码和教学课件等资料。

本书由在计算机程序设计方面有着丰富教学和实践经验的中外作者合作编写。本书是国内首次出版的中英文对照混排式双语版教材的更新版，适合低年级的学生对照阅读，既方便初学者熟悉相关概念和内容，也便于英语非母语的读者熟悉英文专业词汇，尤其适合作为双语教学示范课程的教材。

Paul Kelly是一位治学非常严谨的教师，本书第二作者苏小红在与他合著的过程中，经常与他为一个细节内容的编写反复进行交流与讨论，书稿完成后又进行了多次校对工作。本着对所有读者负责的精神，我们真诚地欢迎读者对教材提出宝贵意见，可以通过发送电子邮件或在相关网站上留言等多种方式与我们交流。

作者的电子邮件地址为paul.kelly@tudublin.ie及sxh@hit.edu.cn。

苏小红
哈尔滨工业大学计算学部

目　　录

Chapter One
Introduction
第 1 章 绪 论

1.1 What is a computer program?
（什么是计算机程序？）

Computers are involved in a wide variety of tasks that we do in our everyday lives. Some of these tasks such as using a word processor or checking e-mail obviously use a computer. Less direct examples occur when we use an ATM at a bank, pay at a supermarket checkout or use a phone.

A computer performs all of these tasks by following a predefined set of instructions. This set of instructions is a called a *computer program*. A computer program to a computer is like a recipe to a chef; it specifies the steps needed to perform a certain task. But unfortunately, unlike a recipe, you can't give your instructions to a computer in a language such as English or Chinese. For instructions to be 'intelligible' to a computer, they need to be expressed in a language 'understood' by the computer. The only language 'understood' by a computer is its own machine language, which consists of a series of binary ones and zeroes.

Machine language is very difficult to use directly and so instructions to a computer are given in a special language called a *programming language*. The programming language is neither English nor machine language, but is somewhere in between. In fact, as you will see, it is more like English than machine language.

Machine languages are known as *low-level language*s and programming languages are known as *high-level languages*.

Writing instructions in a high level language is much easier than writing them in low-level machine language, but is still not as easy as writing them in English or Chinese.

For the computer to carry out the program instructions written in a high level language, they have to be translated from the high level language to the machine language of the computer. A *compiler* does this translation.

现在我们日常生活中的很多工作都要用到计算机。

计算机完成所有这些工作都是通过执行预定义的指令序列来实现的。这个指令序列就称为"计算机程序"。计算机程序和计算机之间，就像食谱和厨师之间的关系一样；计算机程序指定了完成某一任务需要的步骤。

但是遗憾的是，不同于食谱，我们不能用英文或者中文这样的自然语言向计算机发送指令。对于计算机来说，指令必须是"易理解"的，必须表示成一种计算机能"理解"的语言。计算机能"理解"的唯一语言就是机器语言，机器语言由一系列二进制数 0 和 1 组成。

由于机器语言很难直接使用，所以计算机指令被表示成一种特殊的语言，这种特殊的语言称为"程序设计语言"。程序设计语言既不是英语，也不是机器语言，而是一种介于它们两者之间的语言。实际上，正如下面即将看到的那样，较之机器语言，程序设计语言更像英语。

机器语言被认为是一种低级语言，而程序设计语言则被认为是一种高级语言。

为了让计算机执行由高级语言编写的程序指令，必须把这些指令从高级语言形式翻译成计算机的机器语言形式。编译器用于完成这种翻译。

1.2　Developing a computer program
（开发计算机程序）

The first step in developing a computer program is to define and understand the problem to be solved. If you cannot understand the problem then you will certainly not be able to tell a computer how to solve the problem. This is called the analysis phase and basically answers the question "what is to be done?", ignoring for the time being the question of how the problem is to be solved.

Once it is known what has to be done, the question "how is it to be done?" arises. This is called the design phase and it is in this phase that a solution to the problem is developed. The design phase may reveal problems not previously considered in the analysis phase. So rather than being independent phases, the design and analysis phases are closely related and interact with each other.

The analysis and design phases are important to developing a successful solution to a problem. Neglect at either of these phases will result in a 'solution' that will not solve the original problem and may even contribute to making it worse.

Analysis and design are subjects in their own right and are not covered in this book. This book concentrates on the next phase: writing, compiling and testing C++ programs.

There are many different compilers on the market, with updated versions being released regularly. Up to date instructions for writing, compiling and running C++ programs with some of the popular compilers can be found on the web site for this book at http://www. hxedu.com.cn.

1.2.1　Program development cycle

Despite the differences between compilers, the following is a general description of the steps involved in the development of a C++ program.

Step 1: Design the program.

A computer system consists of one or more programs. Each program has to be individually designed to implement the overall solution developed at the analysis and design phases.

After each program is designed, it is important to check its logic before starting to write the program.

Step 2: Write the program.

Firstly, the C++ program instructions are typed into a file using a *text*

开发计算机程序的第一步就是定义和理解要解决的问题。

如果不理解要解决的问题是什么，那么就无法告诉计算机如何去解决这个问题。这个阶段称为分析阶段。分析阶段主要回答"做什么"的问题，而不是考虑如何去做的问题。

一旦知道了要做什么，那么"如何去做"的问题又产生了。这称为设计阶段，在这个阶段要找到一种解决问题的方法。在设计阶段也可能遇到一些在分析阶段没有考虑到的问题。所以，分析和设计不是两个独立的阶段，而是两个密切相关、相互影响的阶段。分析和设计阶段对找到一种成功解决问题的方案来说是十分重要的。忽视这两个阶段中的任何一个都会导致最终的解决方案无法解决最初的问题，甚至还可能产生更坏的结果。

分析和设计属于不同范畴的问题，不在本书的讨论范围之内。本书重点介绍其后面的阶段：C++程序的编写、编译和测试。

尽管编译器之间存在差异，但可以给出开发一个C++程序的一般步骤。

第一步：设计程序。

计算机系统由一个或者多个程序组成。每个程序必须单独设计，以实现在分析阶段和设计阶段提出的整体解决方案。

在程序设计好之后、编写程序之前，检查一下程序的逻辑是很重要的。

第二步：编写程序。

使用文本编辑器把C++程序指

editor. The file containing the C++ statements is called the *source file* and is usually stored on disk. The program instructions are also called the *source code* or the *program code*.

Step 3: Compile the program.

Next the C++ program is passed through a compiler, which translates the C++ program instructions to machine instructions. The compiler reads the source file, translates the C++ statements into machine or *object code*, and stores the object code in an *object file*.

Some errors are likely to occur in this step. An error in the source code is indicated by the compiler and is referred to as a *compile-time error*. The simplest kind of compile-time error is a *syntax error*. This type of error is relatively easy to correct, as the compiler will indicate where in the source code the error has occurred. Typical syntax errors involve missing punctuation and misspellings. All compile-time errors must be corrected before the compilation process can be completed.

The compiler may sometimes issue a warning message during compilation. A warning message is not as serious as a syntax error and will not prevent the program from being compiled. However, warnings are the compiler's way of drawing attention to what it 'thinks' may be a problem and should be investigated.

Step 4: Link the program.

The final step before running the program is to *link* the program using the *linker*. Linking involves combining the object file of the program with other object files from the C++ run-time library to form an *executable file*.

Step 5: Test the program.

When the program is run you may find that it is not working as expected. The fact that a program does not have any compile-time errors does not guarantee that the program will work as required. For example, the programmer may mistakenly have given an instruction to divide a number by zero. This type of error is called a *run-time error* and causes the program to stop before it has completed its task.

The program may complete its task but produce incorrect results or display them at the wrong position on the screen.

These kinds of errors are known as *logic errors* or more commonly as *bugs*. Bugs are much harder to find and fix than compile-time errors.

令输入到一个文件中。这个包含C++语句的文件称为源文件，通常存储在磁盘上。程序指令也称为源代码或程序代码。

第三步：编译程序。

编译器把C++程序指令转换成机器指令。编译器读入源文件，把C++语句翻译成机器代码或目标代码，然后把目标代码存入目标文件中。

在编译阶段很可能出现一些错误。编译器指出的源代码中的错误属于编译时错误。最简单的编译时错误是语法错误。这类错误相对容易纠正，因为编译器会指出这类错误在源代码中的位置。典型的语法错误有遗漏标点和拼写错误。在编译过程完成之前，必须纠正所有的编译错误。

编译器也可能在编译阶段发出一些警告信息。警告信息不像语法错误那样严重，不会妨碍程序的编译。不过，它表示编译器认为可能有问题，希望引起用户的注意并检查一下程序。

第四步：链接程序。

在运行程序之前的最后一步就是使用链接器来链接程序。

所谓链接就是把程序的目标文件和C++运行时库文件结合起来形成一个可执行文件。

第五步：测试程序。

程序在运行时有可能没有像预期的那样工作。一个程序没有编译时错误并不能保证这个程序就能按照要求运行。例如，编程人员可能错误地在一条指令中把 0作为除数。这种错误称为运行时错误，它会导致程序在完成它的任务之前停止。

虽然程序运行完毕，但是结果却是错误的，或者结果显示的屏幕位置是错误的。这类错误称为

Some bugs appear only under certain conditions, for example when a program is run with a particular set of data.

The process of locating and correcting program errors is called *debugging*.

Step 6: Debug the program.

Once a bug has been identified, the next step is to find where in the source code the problem lies. Many compilers have tools that can be used to help locate bugs.

Correcting bugs involves going back to step 2, but hopefully not any further. Going back to step 1 or even back to the analysis and design phases would be like asking an architect to redesign parts of a house while it is being built! In general, try to catch errors as early as possible.

1.3　Learning C++（学习 C++）

You'll learn more from designing, writing, running, and correcting programs than you ever will by simply reading a book. A successful approach to learning to program in C++ depends on large amounts of practice.

To help you practise, there are exercises at the end of each chapter. Do as many of the exercises as possible and get some feedback on your solutions from people who know C++. There are solutions to selected exercises at the web site for this book.

1.4　Web site for this book（本书的网站）

The web site for this book is at http://www.hxedu.com.cn. The source code for all the example programs used in this book as well as answers to selected exercises are available here.

1.5　Brief history of C++（C++ 简史）

C++ is a direct descendent of the C programming language, which was originally developed in 1972 at AT&T Bell Laboratories, New Jersey, USA. C evolved from a language called B, which in turn evolved from a language called BCPL (Basic Combined Programming Language). C++ was developed by AT&T Bjarne Strostrup at Bell Laboratories from 1979 to 1983. The initial version of the language was called "C with Classes" and was used internally in AT&T in 1983. Later that year, the name was changed to "C++". The first commercial version of C++ was marketed in 1985, and since

逻辑错误或者俗称为"bug"（缺陷）。缺陷比编译时错误更难发现和修正。某些 bug 仅在特定的条件下才出现，例如，仅在特定的数据集上运行程序时才出现。定位和修正程序错误的过程称为调试。

第六步：调试程序。

bug 一旦被确认，紧接着就要定位其在源代码中的位置。很多编译器都提供了可用于帮助定位缺陷的工具。

改正一个缺陷后要返回第二步，但是希望不要返回到更前面去。如果返回到第一步甚至是分析和设计阶段，就好像要求建筑师重新设计一座正在建造中的房子一样。一般情况下，应该尽可能早地发现错误。

C++ 语言是由 1972 年诞生于美国新泽西州 AT&T 贝尔实验室的 C 语言直接发展而来的，C 语言是从一种称为 B 的语言演化而来的，这种 B 语言又是从 BCPL 语言演化而来的。

C++ 是 Bjarne Strostrup 于 1979 年至 1983 年在 AT&T 贝尔实验室工作期间完成开发的。该语言的最

then, C++ has evolved to the latest ANSI/ISO (American National Standards Institute/International Organization for Standardization) C++ standard.

1.6　ANSI/ISO C++ standard（ANSI/ISO C++ 标准）

The example programs used in this book conform to the ANSI/ISO C++ standard. Not all compilers conform to this standard, so some compilers may not correctly compile the example programs. Some of the example programs may have to be modified for use with these other compilers. See the web site for details.

初版本称为"有类的C"，并且于1983年在AT&T内部使用。从那以后，该语言被改名为 C++。第一个商业版本的C++在1985年面世。从那以后，C++逐渐发展成为最新的 ANSI/ISO标准。

Chapter Two
Beginning to Program in C++
第 2 章　C++ 编程入门

As in other programming languages, data can be of two types: *constants* and *variables*.

与其他编程语言相同，C++中的数据可分为两种：常量和变量。

2.1　Constants（常量）

As the name suggests, a constant does not change its value in a program. Some examples of constants are shown in Table 2.1 below.

顾名思义，常量的值在程序中是不能改变的。

Table 2.1　Some examples of constants

Type of constant	Examples	Remarks
integer	100, -3, 0	Whole numbers that can be positive, negative or zero.
floating-point	0.34, -12.34, 8.0	Numbers with decimal parts.
character	'x', 'X', '*', '9'	A single character enclosed in single quotation marks.
string	"abc","A100", "9"	One or more characters enclosed in double quotation marks.

integer：整型。

floating-point：浮点型（也称为实型）。

character：字符型。

string：字符串型。

2.2　Variables（变量）

Unlike a constant, a variable can vary its values in a program, and a variable must be defined before it can be used. A variable is defined by giving it a data type and a name.

与常量不同，变量的值在程序中是可以改变的。变量在使用之前必须先定义。定义变量就是指定它的数据类型和名字。

Program Example P2A

```
1   int main( void )
2   {
3     int v1;
4     float v2;
5     char v3;          ←——— Lines 3, 4 and 5 define three variables.
6     v1 = 65;          ←——— Lines 6, 7 and 8 assign a value to each
7     v2 = -18.23;              variable.
8     v3 = 'A';
9     return 0 ;
10  }
```

The line numbers on the left are for reference purposes only and are not part of the program.

C++ programs start with the line

```
int main( void )
```

This marks the point where a C++ program starts to execute and must appear once only in a C++ program.

The program statements are contained within the braces { and } on lines 2 and 10. Each statement ends with a semicolon. The spaces before the semicolon and on each side of the equals sign are not essential and are used here only to improve the readability of the program.

Lines 3, 4 and 5 of this program define three variables: v1, v2, and v3. A variable can be given any name, called an *identifier* in C++, provided it is within the following rules:

- An identifier can only be constructed using letters, numerals or underscores (_).
- An identifier must start with a letter or an underscore.
- An identifier cannot be a C++ *keyword*. A keyword is a word that has a special meaning. (See appendix A for a list of C++ keywords.)
- An identifier can contain any number of characters, but only the first thirty-one characters are significant to the C++ compiler.

Lines 3 to 5 of program P2A define v1 as an integer variable, v2 as a floating-point variable, and v3 as a character variable. Note that identifiers are *case-sensitive*, i.e. the variable V1 is different from the variable v1.

A variable is like a box in the memory of the computer. The box has a name and contains a value. Each box has a name given to it by you, the programmer. The box keeps its value until it is changed by replacing it with some other value.

Lines 6 to 8 of the program assign values to the variables. The value assigned to each variable is stored in the computer's memory.

2.3 Simple output to the screen
（简单的屏幕输出）

Now that values are assigned to the variables, how are the values displayed on the screen? This can be done with cout (pronounced see-out), as demonstrated in the next program.

记号main( )指定了C++程序执行的起点，它在程序中只能出现一次。

程序语句用一对花括号"{}"括起来。每条语句都以分号（;）结尾。分号之前及等号两边的空格不是必需的，增加空格只是为了增强程序的可读性。

变量名是C++的一种标识符，可以给变量起任何名字，只要遵守如下的标识符命名规则即可：

- 标识符只能由字母、数字和下画线（_）组成。
- 标识符必须以字母或下画线开头。
- 标识符不能是C++关键字。关键字是指在程序中具有特殊意义的单词（见附录A中C++关键字的列表）。
- 标识符可以包含任意多个字符，但是C++编译器只识别前31个字符。

注意，标识符是区分大小写的。

变量就像计算机内存中的一个盒子。这个盒子有一个名字和一个值，程序员可以为每个盒子命名。盒子中的数据在程序员给它赋予新值之前是保持不变的。

变量在被赋予新值后，如何将这些新值显示在屏幕上呢？可以用cout（读作c-out）来输出变量的值，如下面程序所示。

Program Example P2B

```
1  #include <iostream>
2  using namespace std ;
3  int main( void )
4  {
5    int v1 ;
6    float v2 ;
7    char v3 ;
8    v1 = 65 ;
9    v2 = -18.23 ;
```

```
10    v3 = 'a' ;
11    cout << "Program Example P2B" << endl ;
12    cout << "v1 has the value " << v1 << endl ;    Display the values
13    cout << "v2 has the value " << v2 << endl ;    of the variables on
14    cout << "v3 has the value " << v3 << endl ;    the screen.
15    cout << "End of program" << endl ;
16    return 0 ;
17  }
```

When you compile and run this program you will get the following on your screen:

```
Program Example P2B
v1 has the value 65
v2 has the value -18.23
v3 has the value a
End of program
```

Line 1 is an example of a *preprocessor directive* and is pronounced "hash include io stream". This line, along with line 2, will be in every program that involves output to the screen or input from the keyboard. The #include preprocessor directive causes the file in the angle brackets, called a *header file*, to be made available to the program. The header file iostream contains some C++ statements to make it easy to perform input and output from a C++ program. (Preprocessor directives are covered in Appendix G.)

C++ refers to your computer as a console. To display data to the screen, C++ sends the data to cout (console output). cout is called an output stream object.

In C++, letters or numbers enclosed in double quotation marks is called a string of characters or a character string.

In line 11 the stream insertion operator << is used to insert a string of characters into the output stream object cout. endl (end line, pronounced end-ell) is used to go the start of the next line on the screen.

程序的第1行是一个编译预处理命令。这一行和下一行将会出现在每一个需要向屏幕输出数据或者从键盘输入数据的程序中。位于尖括号内的文件称为头文件，编译预处理命令#include可使头文件在程序中生效。头文件iostream中包含一些帮助程序轻松实现数据输入/输出的C++语句（编译预处理命令将在附录G中详细介绍）。

C++语言将计算机看作一个控制台。要想在屏幕上显示数据，只要将数据送到称为标准输出流对象的cout（控制台输出）即可。

在 C++语言中，用一对双引号括起来的字符或数字，称为字符序列或字符串。

在程序的第11行中，流插入运算符<<用于向标准输出流对象cout输出字符串。endl（行结束）用于在屏幕上实现光标换行的功能。

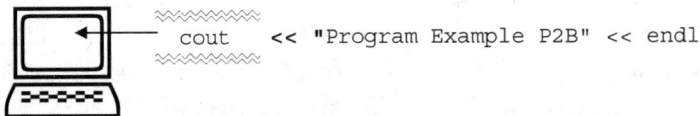

```
cout << "Program Example P2B" << endl
```

Line 12 has a string of characters enclosed in double quotation marks followed by << and a variable name. The variable name is not in quotation marks.

This statement displays the string of characters followed by the value of the variable v1.

The variable v1 has the value 65, so the output from line 12 is:

```
v1 has the value 65
```

程序的第12行中有一个用双引号括起来的字符串，后接<<和变量名。变量名无须用圆括号括起来。

The spaces on either side of << on line 12 are for readability only. The stream insertion operator << may be used any number of times with cout. For example, you can combine lines 14 and 15 as:

```
cout<<"v3 has the value "<<v3<<endl<<"End of program"<<endl;
```

If the line gets too long it can split up between two or more lines:

```
cout<<"v3 has the value "<<v3<<endl
    <<"End of program"<<endl;
```

2.4 Comments（注释）

Comments are added to a C++ program to make it more readable for the programmer, but they are completely ignored by the compiler. In C++, comments start with the characters //. Some comments can be added to the last program.

在C++程序中增加注释可使程序可读性更好，而编译器则完全忽略注释。在 C++中，注释以 // 开头。

Program Example P2C

```
1   // Program Example P2C
2   // Program to introduce variables in C++.
3   #include <iostream>
4   using namespace std ;
5
6   int main( void )
7   {
8     int v1 ;     // v1 is an integer variable.
9     float v2 ;   // v2 is a floating-point variable.
10    char v3 ;    // v3 is a character variable.
11
12    // Now assign some values to the variables.
13    v1 = 65 ;
14    v2 = -18.23 ;
15    v3 = 'a' ;
16
17    // Finally display the variable values on the screen.
18    cout << "Program Example P2C" << endl ;
19    cout << "v1 has the value " << v1 << endl ;
20    cout << "v2 has the value " << v2 << endl ;
21    cout << "v3 has the value " << v3 << endl ;
22    cout << "End of program" << endl ;
23    return 0 ;
24  }
```

Comments can be placed anywhere in a C++ program and start after // and end at the end of the line. C++ comments cannot span more than one line.

The older C-style comments may also be used in a C++ program. Comments in the C programming language begin with the characters /* and end with the characters */. Unlike C++ comments, C-style comments can span more than one line.

注释可以放在C++程序的任何地方，它必须以//开始，到本行末尾结束，且只能占一行。
在C++程序中，也可以使用传统的C风格的注释，即以/*开头，到*/结束。与C++风格的注释不同,C风格的注释可以跨越多行。

Typically, comments are placed at the start of the program to describe the purpose of the program, the author, date written and any other relevant information, such as the version number. For example:

通常，在程序开头都应该写一段注释，用于描述程序的功能、编程人员姓名、编程日期及其他相关信息，如版本号等。

```
//    Program name  : P2A.
//                    Introduction to variables.
//    Written by    : Paul Kelly and Su Xiaohong
//    Date          : 29/6/2023.
//    Version number: 3.0
```

You can also write these comments in the C-style:

```
/*    Program name  : P2A.
                      Introduction to variables.
      Written by    : Paul Kelly and Su Xiaohong
      Date          : 29/6/2023.
      Version number: 3.0                         */
```

Comments are also used to describe in plain language the function of a particular section of a program. Get into the habit of using comments; the more complicated the program becomes the more valuable they are to the programmer. However, writing bad comments can lead to confusion and are worse than no comments at all. A comment must be easier to understand than the code that it is trying to explain!

注释的目的就是使用简单易懂的语言来描述程序中一个特定部分的功能。应该养成书写注释的习惯。对于程序员来说，程序越复杂，注释就越有价值。但是，不好的注释也会把人弄糊涂，甚至还不如不写注释。注释必须比程序代码更容易让人理解！

The blank lines at 5, 11 at 16 are used to separate different sections of the program. A blank line can be placed anywhere in the program and like comments are ignored by the compiler.

2.5　Data types（数据类型）

In previous programs, it was show how to declare a variable and associate it with a particular data type (char, int or float). The C++ language has a variety of other data types besides the three basic types of char, int and float. Different data types require different amounts of memory and therefore vary in the range of values they can store.

在前面的程序中，我们已经学习了怎样声明一个变量，并将它和一个具体的数据类型（char、int 和float）联系起来。除了char、int 和float这三种基本数据类型，C++还提供了很多其他数据类型。不同的数据类型占用不同大小的内存空间，因此，它们所能表示的数据的范围也各不相同。

Details of the various C++ data types are given in appendix D.

2.5.1　Short integer data types

To define a variable v1 as a short integer:

```
short int v1 ;
```

short integer：短整型。

The keyword int is optional here, so v1 can also be defined as:

```
short v1 ;
```

2.5.2　Long integer data types

To define a variable v2 as a long integer:

```
long int v2 ;
```

The keyword `int` is optional here, so `v2` can also be defined as:

```
long v2 ;
```

2.5.3   Boolean data types

The Boolean data type `bool` can store only one of two values: `true` or `false`. Normally true is numerically 1 and false is numerically 0. The following defines a Boolean variable `v3`:

布尔数据类型（bool）只有true（真）或false（假）这两个值。通常，用数值1表示真，用数值0表示假。

```
bool v3 ;
```

2.5.4   Double floating-point data types

The `double` data type allows you to increase the range and precision (or accuracy) of a floating-point number.
To define a variable `v4` as a `double` data type:

双精度浮点型（double）能够增加浮点型数据表示数的范围及精度。

```
double v4 ;
```

2.5.5   Unsigned integer data types

The keyword `unsigned` extends the positive range of an integer variable but does not allow negative values to be stored.
The following statement defines an unsigned integer variable `v5`:

关键字unsigned扩展了整型变量的正数范围，但是不允许其存储负值。

```
unsigned int v5 ;
```

Appendix D contains a full list of the C++ data types.

附录D列出了C++的全部数据类型。

2.6   Data type sizes（数据类型的大小）

The next program uses the `sizeof` operator to display the number of bytes of memory required by some of the common data types.

下面的程序使用sizeof 运算符来显示一些常见的数据类型在内存中占用的字节数。

Program Example P2D

```
1  // Program Example P2D
2  // Program to display the memory required by C++ data types.
3  #include <iostream>
4  using namespace std ;
5
6  int main( void )
7  {
8    cout << " Data type    Number of bytes" << endl ;
9    cout << " ---------    ---------------"      << endl ;
10   cout << "  char         " << sizeof( char )   << endl ;
11   cout << "  int          " << sizeof( int )    << endl ;
12   cout << "  short int    " << sizeof( short )  << endl ;
13   cout << "  long int     " << sizeof( long )   << endl ;
14   cout << "  bool         " << sizeof( bool )   << endl ;
15   cout << "  float        " << sizeof( float )  << endl ;
16   cout << "  double       " << sizeof( double ) << endl ;
17   return 0 ;
18 }
```

The output from this program is:

```
Data type        Number of bytes
---------        ---------------
 char                 1
 int                  4
 short int            2
 long int             4
 bool                 1
 float                4
 double               8
```

2.7   Operators（运算符）

2.7.1   The assignment operator

The assignment operator (=) is used to assign values to variables. For example, the statement

赋值运算符（=）用于给变量赋值。

```
v = 1 ;
```

assigns the value 1 to the variable v.

Further examples are:

```
total = 0 ;
value = 100.12 ;
reply = 'y' ;
v1 = v2 = v3 = 123 ;
```

In the last example, the value 123 is assigned to the three variables v1, v2, and v3.

2.7.2   Arithmetic operators

There are five arithmetic operators in C ++ as shown in Table 2.2.

Table 2.2   Five arithmetic operators in C ++

Operator	Used for
+	addition
–	subtraction
*	multiplication
/	division
%	modulus (this gives the remainder after division)

The next program demonstrates the use of the arithmetic operators.

Program Example P2E

```
1  // Program Example P2E
2  // Demonstration of the arithmetic operators.
3  #include <iostream>
4  using namespace std ;
5
6  int main( void )
7  {
8      // Define the variables used in the program.
```

```
9      int var1, var2 ;
10
11     // Place values into the variables and display
12     // the values in the variables.
13     var1 = 0 ;
14     var2 = 10 ;
15     cout << "var1 starts at " << var1 << endl
16          << "var2 starts at " << var2 << endl ;
17
18     // Do some arithmetic with the variables and display
19     // the values in the variables.
20
21     var2 = var1 + 18 ;
22     cout << "var2 is now " << var2 << endl ;
23
24     var1 = var2 * 3 ;
25     cout << "var1 is now " << var1 << endl ;
26
27     var1 = var2 / 3 ;
28     cout << "var1 is now " << var1 << endl ;
29
30     var2 = var1 - 1 ;
31     cout << "var2 is now " << var2 << endl ;
32
33     var1 = var2 % 3 ;
34     cout << "var1 is now " << var1 << endl ;
35
36     var1 = var1 + 1;
37     cout << "var1 is finally " << var1 << endl ;
38
39     var2 = var2 * 5 ;
40     cout << "and var2 is finally " << var2 << endl ;
41     return 0 ;
42  }
```

When you run this program the following will be displayed on the screen:

```
var1 starts at 0
var2 starts at 10
var2 is now 18
var1 is now 54
var1 is now 6
var2 is now 5
var1 is now 2
var1 is finally 3
and var2 is finally 25
```

2.7.3   Increment and decrement operators

It is very common in programming to add or subtract 1 from a variable; so common, in fact, that C++ provides operators specifically to do these tasks.

在程序中，常常需要对变量进行增 1 或减 1 操作。C++ 提供了专门进行这类操作的运算符。

In line 36 of program P2E, 1 is added to `var1` by the statement

```
var1 = var1 + 1 ;
```

This statement can be replaced by the statement

```
var1++ ;
```

The increment operator ++ adds 1 to the value of a variable.
The statement

自增运算符（++）可使变量的值增1。

```
var1 = var1 - 1 ;
```

subtracts 1 from the variable `var1`. An equivalent statement is

```
var1 -- ;
```

The next program demonstrates the increment and decrement operators.

自减运算符（--）可使变量的值减1。

Program Example P2F

```
1   // Program Example P2F
2   // Demonstration of the increment and decrement operators.
3   #include <iostream>
4   using namespace std ;
5
6   int main( void )
7   {
8    // Define two variables and initialise them.
9     int var1 = 1, var2 = 2 ;
10
11     cout << "Initial values: " ;
12     cout << "var1 is " << var1
13         << " and var2 is " << var2 << endl ;
14
15     var1++ ;   // Add 1 to var1.
16     var2-- ;   // Subtract 1 from var2.
17
18     cout << "Final values: " ;
19     cout << "var1 is " << var1
20         << " and var2 is " << var2 << endl ;
21     return 0 ;
22   }
```

The output from this program is:

```
Initial values: var1 is 1 and var2 is 2
Final values: var1 is 2 and var2 is 1
```

Line 9 of this program defines the variables `var1` and `var2` as integers and initialises them to 1 and 2, respectively.

Line 15 adds 1 to `var1`, and line 16 subtracts 1 from `var2`.

Lines 19 and 20 display the final value of `var1` and `var2`.

The increment operator ++ has two forms, *prefix* and *postfix*, which are demonstrated in the next program.

自增运算符（++）有两种形式：前缀形式和后缀形式。

Program Example P2G

```
1   // Program Example P2G
2   // Demonstration of the prefix and postfix ++ operators.
3   #include <iostream>
4   using namespace std ;
5
6   int main( void )
7   {
8     int var1, var2, var3, var4 ;
9
10    var1 = var2 = 1 ;
11
12    var3 = var1++ ;  // var3 is 1, var1 is 2.
13    var4 = ++var2 ;  // var4 is 2, var2 is 2.
14
15    cout << "var1 is " << var1
16         << ", var2 is " << var2 << endl ;
17    cout << "var3 is " << var3
18         << ", var4 is " << var4 << endl ;
19    return 0 ;
20  }
```

The output from this program is:

```
var1 is 2, var2 is 2
var3 is 1, var4 is 2
```

The program starts by initialising both variables var1 and var2 to 1. In line 12 the variable var3 is assigned the value of var1 (=1), and then var1 is incremented to 2. This is an example of a *postfix* operation. In contrast, line 13 shows an example of a *prefix* operation. In line 13, var2 is incremented first, and then the new value (2) is assigned to var4.

If the ++ is before a variable, the variable is incremented before it is used. If the ++ is after the variable, the variable is used and then incremented. The difference between prefix and postfix is only relevant where an assignment is involved. So in line 15 of program P2F, either var1++ or ++var1 could be used.

The decrement operator -- also has prefix and postfix forms. If the -- is before a variable, the variable is decremented before it is used. If -- is after the variable, the variable is used and then decremented. Again this is only relevant if an assignment is involved. In line 16 of program P2F either var2-- or --var2 could be used.

2.7.4 Combined assignment operators

A statement such as

```
var = var + 3 ;
```

may also be written as

程序开头将变量var1和var2分别初始化为1。第12行将变量var1的值（1）赋给变量var3，然后将变量var1的值增1，变成2，这是自增运算符作为后缀的一个例子。相比之下，第13行是"++作为前缀"的一个例子。在第13行，先将var2的值增1，然后，将var2的新值（2）赋给变量var4。

如果++放在变量之前，那么变量的值在使用前先增1；如果++放在变量之后，那么变量的值先被使用，然后再将其增1。++作为前缀与作为后缀之所以有所不同，是因为语句中包含了赋值操作，所以，在程序P2F的第15行，使用var1++或者++var1的结果是相同的。

自减运算符也有前缀和后缀两种形式，如果将--放在变量之前，那么先将变量的值减1，然后再使用它。如果将--放在变量之后，那么先使用变量的值，然后再将其减1。同样，--运算符作为前缀与作为后缀之所以有所不同，也是因为语句中包含了赋值操作。因此，在程序P2F的第16行，既可以使用var2--，也可以使用--var2。

```
var += 3 ;
```

The += operator adds the value on its right to the variable on its left. There are five combined assignment operators: +=, -=, *=, /=, and %=, corresponding to the five arithmetic operators +, -, *, /, and %. Table 2.3 shows some examples of their use.

Table 2.3   Some examples of using combined assignment operators

Operator	Examples	Equivalent
+=	count += 11 ;	count = count + 11 ;
	a += b ;	a = a + b ;
-=	count -= 20 ;	count = count - 20 ;
	a -= b ;	a = a - b ;
*=	rabbits *= 2 ;	rabbits = rabbits * 2 ;
	a *= b ;	a = a * b ;
/=	money /= 2 ;	money = money / 2 ;
	a /= b ;	a = a / b;
%=	pence %= 100 ;	pence = pence % 100 ;
	a %= b ;	a = a % b ;

2.8   Operator precedence（运算符的优先级）

Consider the following statement:

```
var = 2 + 7 * 8 ;
```

Does this mean

(a) that 2 is added to 7, giving 9, which is multiplied by 8, giving var a value of 72,

or

(b) that 7 is multiplied by 8, giving 56, which is added to 2, giving var a value of 58?

Clearly the order of evaluation is important. With (a) you get 72 and with (b) you get 58.

C++ has rules to remove any ambiguity present in a statement such as the one above. *Operator precedence*, as shown in Table 2.4, provides these rules.

下面的语句表示(a)和(b)中的哪一种含义呢？

(a) 2与7相加得到9，然后再与8相乘，最后将乘积72赋值给变量var。

(b) 7与8相乘得到56，再与2相加，最后将58赋值给变量var。

显然，运算的先后顺序是至关重要的。(a)得到72而(b)得到58。C++提供了一些规则用于消除上述语句的二义性。

Table 2.4   Operator precedence in C++

Operator	Precedence	Meaning	Associativity
-	Highest	Unary minus	Right to left
*		Multiplication,	
/	Lower	division, and	Left to right
%		modulus	
+	Lowest	Addition and	Left to right
-		subtraction	

From this table, multiplication and division have a higher priority than addition and subtraction, so these operators get evaluated first. Thus the statement

```
var = 2 + 7 * 8 ;
```

从表中可以看出，"*"和"/"的优先级高于"+"和"-"，所以应先进行"*"和"/"运算。

evaluates to

```
var = 2 + 56 = 58.
```

You can use parentheses to change the order of evaluation. Thus the statement

```
var = ( 2 + 7 ) * 8 ;
```

evaluates to

```
var = 9 * 8 = 72
```

当然，也可以使用圆括号改变运算的先后顺序。在任何表达式中，都优先计算圆括号内表达式的值。

because any expressions contained within parentheses get evaluated first. Expressions containing operators of the same precedence are evaluated according to their *associativity*, as shown in table 2.4 above. Associativity gives the order in which operators of the same precedence are evaluated. For example, in the statement

```
var = 1 + 6 * 9 % 5 / 2 ;
```

the *, / and % have equal precedence, so which is done first? The *, / and % associate left to right, so the order is:

```
1 + 6 * 9 % 5 / 2 = 1 + 54 % 5 / 2 = 1 + 4 / 2 = 1 + 2 = 3
```

当表达式中所含运算符的优先级相同时，应根据如表2.4所示的运算符的结合性进行计算。结合性规定了在运算符优先级相同时表达式的计算顺序。

Precedence and associativity can be a source of errors. Use parentheses to group the variables, constants and operators in a clear and unambiguous way. Expressions within parentheses are evaluated first; therefore, using parentheses will remove any confusion about which operations are done first. For example, the last statement could be written much more clearly as:

```
var = 1 + ( ( 6 * 9 ) % 5 ) / 2 ;
```

and the result will be the same.

The unary minus operator, like the binary minus operator, is represented as -.

In the statement

```
var = -3*2-1 ;
```

误用运算符的优先级和结合性往往会导致程序错误，清晰而明确地用圆括号将变量、常量和运算符括起来，可以避免这一问题。由于要优先计算圆括号内的表达式，因此，使用圆括号可以避免在优先计算哪个操作上的混淆。

一元减运算符与二元减运算符的表示形式相同，均为"-"。

the first - is a unary minus and the second - is a binary minus. The unary minus appears before an operand and the binary minus appears between two operands. Because the unary minus has highest priority, the above statement is equivalent to

```
var = ((-3)*2)-1 ;
```

and not

```
var = -(3*2-1) ;
```

一元减运算符放在操作数的前面，而二元减运算符则放在两个操作数的中间。

In the first case var is assigned a value of -7, while in the second case -5 is assigned to var.

2.9　Data type conversions and casts（类型转换和强转）

Consider the following program, which divides an integer variable `var1` by a floating-point variable `var2`, placing the result in a floating-point variable `var3`.

Program Example P2H

```
1   // Program Example P2H
2   // Demonstration of a mixed data type expression.
3   #include <iostream>
4   using namespace std ;
5
6   int main( void )
7   {
8       int var1 = 10 ;
9       float var2 = 2.4 ;
10      float var3 ;
11      int var4 ;
12      float var5 ;
13
14      var3 = var1 / var2 ; // Mixed expression assigned to a
15                           // float.
16      var4 = var1 / var2 ; // Mixed expression assigned to an
17                           // int.
18      var5 = var1 / 4 ;    // Non-mixed expression assigned to a
19                           // float.
20      cout << "var3 = " << var3
21          << " var4 = " << var4
22          << " var5 = " << var5 << endl ;
23      return 0 ;
24  }
```

The output from this program is:

```
var3 = 4.16667 var4 = 4 var5 = 2
```

Line 14 is an example of a *mixed expression*, where the variable `var1` of type `int` is divided by the variable `var2` of type `float`. The value of the variable `var1` is automatically converted to a `float` before being divided by the value of `var2`. The result of the division is a value of type `float`, which is then assigned to the `float` variable `var3`. (Note that only a copy of the value of `var1` is converted to a `float`. The variable `var1` is still an `int`.)

In line 16, the result of the division is assigned to an `int` type variable `var4`, resulting in a loss of the fractional part of the number.

In line 18, the division involves two integers, so there is no need for any type conversions. The result of the division is the integer value 2, which is converted to a `float` and assigned to `var5`.

When doing calculations involving mixed data types, C++ ranks the data types in this order:

程序的第14行是一个混合类型表达式运算的例子，类型为int的变量var1除以类型为float的变量var2。变量var1在被var2除之前，其值会被自动转换成float类型，除法运算的结果为float类型，它将被赋值给类型为float的变量var3。（注意，只是将var1值的副本转换为float类型，而变量var1的类型仍然是int。）

在第16行，除法运算的结果被赋值给整型变量var4时，其小数部分将被截断。

在第18行，被除数与除数都是整型变量，所以不需要类型转换，

```
char < short < int ≤ long < float < double
```

and a signed data type is less than the corresponding unsigned data type (for example, a `signed int < unsigned int`).

For calculations involving mixed data types, C++ automatically converts the value in the lower data type to a higher type. This is called a *promotion* or *widening* of the data. Promotion will cause no loss of data, because the higher data types occupy more memory than the lower types and can therefore hold the data precisely. On the other hand, when data is assigned to a variable of a lower type, *demotion* or *narrowing* occurs. Demotion may result in a loss of data, because the lower data type may not have enough bytes of memory to store the higher data type. In addition to automatic conversion, C++ allows you to perform manual conversion with a *static cast*. The general format of a static cast is:

```
static_cast<type>( expression )
```

Using a static cast is equivalent to assigning the expression to a variable of the specified type and using that variable in place of the expression. For example, if you change line 18 of P2H to

```
var5 = static_cast<double>( var1 ) / 4 ;
```

the value of `var1` is cast from an integer to a double floating-point value. The expression is now mixed, and the integer value 4 will be promoted to a `double`. The result of the division is now 2.5, which is assigned to `var5`. Without the static cast, `var5` would have a value of 2.0. The same result is achieved by rewriting line 18 as:

```
var5 = var1 / 4.0 ;
```

Here `4.0` is a floating-point constant. The expression `var1/4.0` is now mixed, and the value of `var1` is therefore promoted to a floatingpoint value automatically.

As a further example of static casting, consider two `double` variables defined as

```
double num1 = 1.9, num2 = 2.9 ;
```

To display the total of the integer parts of these numbers, the following statement can be used:

```
cout << static_cast<int>( num1 ) + static_cast<int>( num2 ) ;
```

the values in `num1` and `num2` are cast to integers, losing their fractional parts and resulting in the total of 3 being displayed.

Note: `(float)(expression)` and `float(expression)` are older equivalents of `static_cast<float>(expression)`. These may still be used, but the newer `static_cast` is recommended.

运算结果仍然为整型值（即2），最后在将其赋值给浮点型变量var5时，将其转换为float类型。

在进行混合类型运算时，C++自动将数据从类型级别较低的一方向类型级别较高的一方转换，这称为数据类型的"提升"或"扩展"。因为相对于级别较低的数据类型，级别较高的数据类型要占用更大的内存空间，因此，类型提升后不会导致数据信息的丢失，从而能够保存更精确的数据。另外，将数据赋值给类型级别较低的变量时，就会发生类型的"降级"和"收缩"。由于级别较低的类型没有足够的存储空间用来存储级别较高的数据，因此类型的降级会导致数据信息的丢失。

除了自动类型转换，C++还允许使用强制类型转换运算符进行人为的类型转换。

Programming pitfalls

1. Do not type a semicolon after either

```
#include <iostream>
```

or

```
main()
```

2. End each C++ statement with a semicolon.
3. A semicolon is not always at the end of a line. For example,

```
int x, y // x and y coordinates of a point ;
```

This will cause a compiler error, because the semicolon is part of the comment.

4. A typing error may result in a statement that does nothing but that is valid nonetheless.

For example,

```
a + b ; // Valid, but the result is not stored.
```

This was probably meant to be:

```
a += b ; // Adds b to a and stores the result in a.
```

or:

```
a = b ; // Assigns b to a.
```

5. A variable must have a value before using it in an arithmetic expression.

```
int counter ; // What value is in counter?
counter++ ;
```

These statements add 1 to the value in the storage location occupied by counter. We have no idea what this value is, as the program did not assign counter a value. The problem is fixed by initialising counter to some value.

For example:

```
int counter = 0 ;
```

6. Be aware of the operator precedence rules. If you are unsure, use parentheses. If you are using parentheses in an expression, count the number of opening and closing parentheses. They should be equal.

7. Each variable has an associated data type (int, float, etc.). Be careful not to go outside the range of valid values for a variable. For example, a short int cannot hold values bigger than 32,767 or smaller than −32,768. See appendix D.

1. 不要在下面语句的行末添加分号。

2. 除了上述语句，C++中的每条语句都应以分号结束。
3. 分号并非总是在行末出现。

4. 一个录入错误可能导致一条语句什么都没有做，但它却是有效的。

5. 在算术表达式中使用变量前，必须将其初始化。

6. 一定要了解运算符的优先级。如果不能确定运算符的优先级，那么应该使用圆括号。使用圆括号时，一定要注意圆括号是否配对，左圆括号和右圆括号的个数一定要相等。

7. 每个变量都有一个关联的数据类型（int，float，等等）。一定不要让变量的值超出变量所

8. You may not always get what you expect when doing arithmetic in C++.

   For example:

```
int a = 100 ;
int b = 8 ;
float f ;

f = a / b ;
```

   The variable f will contain 12, and not the 12.5 you would expect. As a and b are integers, integer arithmetic is performed, resulting in the loss of the fractional part of the result. Use a static cast if the fractional part of the result is required. For example:

```
f = static_cast<float>( a ) / b ;
```

9. Avoid unnecessary comments in your program. For example,

```
int a ; // a is an integer
a = 1 ; // assign 1 to a
```

   Both of the above comments are unnecessary.

能表示的范围。例如，short int 类型不能存储大于32 767或小于−32 768的整数。

8. 在C++算术运算中，期望的结果不一定总能得到。

9. 避免不必要的、无意义的注释。

Quick syntax reference

At the end of each chapter the most important features of the C++ syntax covered in the text are briefly summarised. While not covering the strict definition of the syntax, which can be complex for a beginner, it should prove to be a useful "memory jog" while writing programs.

	Syntax	Examples

Start of program	`#include <iostream>`
	`using namespace std ;`
	`main()`
	`{`	

Defining variables	`char variables ;`
	`int variables ;`
	`short int variables ;`
	`long int variables ;`
	`float variables ;`
	`double variables ;`
	`unsigned variables ;`
		`char grade, reply ;`
		`int number ;`
		`short int exam_mark ;`
		`long int light_years ;`
		`float average, percentage ;`
		`double number, remainder ;`
		`unsigned int student_number ;`

Assignment	`=`	`number = 59.75 ;`

Comments	`//`
	or
	`/*    */`
		`// A one line comment.`
		`/* A comment split`
		`    over`
		`    more than one line. */`

Arithmetic operators	`+`
	`-`
	`*`
	`/`
	`%`
		`tax = sales * 0.21 ;`
		
		`remainder = number % 10 ;`
		
		`average = ( n1 + n2 ) / 2 ;`

(cont.)

	Syntax	Examples
Display to screen	cout <<	cout << "tax = " << tax ;
Increment and decrement	++ --	int n1 = 10, n2 = 20, n3, n4 ; n1++ ; n2-- ; n3 = ++n1 ; n4 = n2-- ;
Combined operators	+= -= *= /= %=	n1 += 10 ; n2 *= n1 ;
Casting	static_cast<type>	int n ; static_cast<float>(n)
End of program	}	

Exercises

1. Which of the following are valid C++ variable names? If valid, do you think the name is a good mnemonic (i.e. reminds you of its purpose)?

    (a) `stock_code`

    (b) `money$`

    (c) `Jan_Sales`

    (d) `X-RAY`

    (e) `int`

    (f) `xyz`

    (g) `1a`

    (h) `invoice_total`

    (i) `John's_exam_mark`

2. Which of the following are valid variable definitions?

    (a) `integer account_code ;`

    (b) `float balance ;`

    (c) `decimal total ;`

    (d) `int age ;`

    (e) `double int ;`

3. Write variable definitions for each of the following:

    (a) integer variables `number_of_transactions` and `age_in_years`

    (b) floating-point variables `total_pay`, `tax_payment`, `distance` and `average`

    (c) long integer variables `record_position` and `count`

    (d) a character variable `account_type`

    (e) a double variable `gross_pay`

4. Write a variable definition for each of the following:

    (a) a number of students

    (b) an average price

(c) the number of days since 1 January 1900

(d) a percentage interest rate.

5. Assuming the following,

```
int i ;
char c ;
```

which of the following are valid C++ statements?

```
c = 'A' ;
i = "1" ;
i = 1 ;
c = "A" ;
c = '1';
```

6. Write a C++ program to assign values to the variables in exercise 3 and display the value of each variable on a separate line.

7. Write a C++ program to display your name and address on separate lines.

8. Convert the following mathematical equations into valid C++ statements:

(a) $m = \dfrac{y_1 - y_2}{x_1 - x_2}$

(b) $y = mx + c$

(c) $a = \dfrac{b}{c} - \dfrac{d}{e}$

(d) $C = \dfrac{5(F - 32)}{9}$

(e) $s = ut + \frac{1}{2}at^2$

9. Assuming the following variable definitions,

```
int a = 1, b = 10, c = 5 ;
int d ;
```

what is the value of d after each of the following statements?

(a) `d = b / c + 1 ;`

(b) `d = b % 3 ;`

(c) `d = b - 3 * c / 5 ;`

(d) `d = b * 10 + c - a * 5 ;`

(e) `d = ( a + b - 1 ) / c ;`

(f) `d = ( ( -a % c ) + b ) * c ;`

(g) `d = --a ;`

10. Assuming the same variable definitions as in exercise 9, correct the errors in the following C++ statements:

(a) `d = 2 (b + C) ;`

(b) `d = 5b + 9c ;`

(c) `d = b - 3 X 19 ;`

(d) `d = b.c + 10 ;`

(e) `d = ( a + b ) / c ;`

11. Write suitable statements to perform the following:

    (a) add 1 to num1, placing the result in num1

    (b) add 2 to num1, placing the result in num2

    (c) add 2 to num2, placing the result in num2

    (d) subtract 1 from num1, placing the result in num1

    (e) subtract 2 from num2, placing the result in num2

12. Assuming the following,

    ```
    int a = 12, b = 0, c = 3, d ;
    ```

    what is the value of a, b, c and d after each of the following statements?

    (a) a++ ;

    (b) b-- ;

    (c) d = ++c ;

    (d) d = c-- ;

    (e) d = a++ - 2 ;

    (f) d = a++ + b++ - c-- ;

13. Assuming the following,

    ```
    int a = 1, b = 2, c = 3 ;
    ```

    what is the value of a, b and c after each of the following statements?

    (a) a += b ;

    (b) a /= 3 ;

    (c) a *= c ;

    (d) a %= 2 ;

    (e) a += b+1 ;

    (f) a += ++b ;

14. Place parentheses around the following expressions to indicate the order of evaluation as given in the C++ operator precedence table in appendix B.

    (a) a = 1 - 2 * 3 + 4 / 5 ;

    (b) a = 5 % b + c - d / 10 ;

    (c) a = ++b * -10 / 5 ;

15. Assuming the following,

    ```
    char ch_val ; int int_val ; long long_val ;
    float float_val ; double double_val ;
    unsigned int unsigned_int_val ;
    ```

    which of the following may lose data because of demotion?

    (a) int_val = long_val ;

    (b) int_val = ch_val ;

    (c) double_val = float_val ;

    (d) long_val = float_val ;

    (e) int_val = unsigned_int_val ;

    (demotion：类型降级)

16. Assuming the following,

```
int a = 5, b = 4 ;
float c = 3, d ;
```

what is the value of d after each of the following?

(a) `d = a / b ;`

(b) `d = static_cast<float>( a ) / b ;`

(c) `d = c / b ;`

(d) `d = static_cast<int>( c ) / b ;`

(e) `d = a / 2 ;`

(f) `d = a / 2.0 ;`

(g) `d = static_cast<float>( a ) / 2 ;`

(h) `d = static_cast<int>( c ) % 2 ;`

17. The following statements are supposed to swap the values in the integer variables a and b. Do they? Assume a is 1 and b is 2.

```
a = b ;
b = a ;
```

18. Write a program to compute

(a) the volume and

(b) the surface area

of a box with a height of 10 cms, a length of 11.5 cms, and a width of 2.5 cms.

(volume : 体积 ; surface area : 表面积)

19. Given the following variables:

```
double purchase_price, selling_price, profit, percentage_profit ;
```

Write a program to

(a) assign a value of fifty Euro to `purchase_price`

(b) assign a value of sixty Euro to `selling_price`

(c) assign to `profit` the value of the difference between `selling_price` and `purchase_price`

(d) assign to `percentage_profit` 100 times the value of `profit` divided by `purchase_price`

(e) display the value of all four variables.

20. Write a program to do the following:

(a) calculate and display the sum of the integers 1 to 5

(b) calculate the average of the floating-point numbers 1.0, 1.1, 1.2,..., 2.0.

Chapter Three
Keyboard Input and Screen Output
第 3 章　键盘输入和屏幕输出

In C++, data stream objects are used to perform basic input and output of data to and from various devices such as the keyboard and the screen. A stream is a data communication object connected to an input or output device.

Just as standard output stream cout is automatically associated with the screen, the standard input stream cin (console input) is automatically associated with the keyboard. The insertion operator << is used to write data to cout and the extraction operator >> is used to read data from the keyboard.

Essentially, reading data from the keyboard is the opposite of writing data out to the screen. The opposite of cout is cin and the opposite of << is >>.

在C++中，数据流对象用于在各种不同的设备（如键盘和屏幕）上执行基本的数据输入和输出操作。所谓流，就是与输入/输出设备相关联的数据通信对象。

正如标准输出流cout自动与屏幕相关联一样，标准输入流cin自动与键盘相关联。流插入运算符<<用于将数据输出到cout，流提取运算符>>用于从键盘读取数据。

3.1　Simple keyboard input（简单的键盘输入）

The following program reads a number from the keyboard and stores it in the variable num (for number).

下面的程序从键盘读入一个数，并将其保存到变量num中。

Program Example P3A

```
1   // Program Example P3A
2   // Program to demonstrate keyboard input.
3   #include <iostream>
4   using namespace std ;
5
6   int main( void )
7   {
8     int num ;
9
10    cout << "Please type a number: " ;
11    cin >> num ;
12    cout << "The number you typed was " << num << endl ;
13    return 0 ;
14  }
```

This program simply asks the user for a number and displays the entered number on the screen.

Line 10 displays the message:

```
Please type a number:
```

Line 11 causes the computer to wait indefinitely until you type a number and press the Enter key.

When you type in a number (e.g. 123) followed by the Enter key, the program continues to line 12 and displays:

```
The number you typed was 123
```

As a memory aid when using `cin`, think of the data as entering the input stream and flowing in the direction indicated by the arrows >> i.e. towards the memory variable.

使用cin时，可以认为数据先进入输入流中，然后按照箭头>>所指的方向流入内存中的变量。

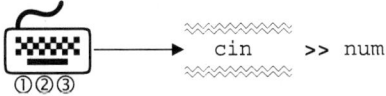

```
cin    >> num
```

Similarly when using cout, think of the data as entering the output stream and flowing in the direction indicated by the arrows << i.e. towards the screen.

类似地，使用cout时，可以认为数据先进入输出流中，然后按照箭头<<所指的方向送到屏幕显示。

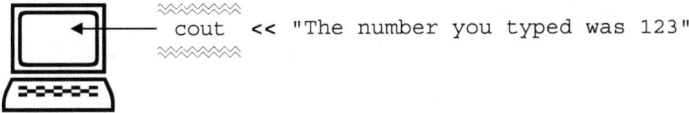

```
cout  << "The number you typed was 123"
```

The next program inputs two floating-point numbers from the keyboard and displays the result of their addition.

Program Example P3B

```
1   // Program Example P3B
2   // Program to input two numbers and display their sum.
3   #include <iostream>
4   using namespace std ;
5
6   int main( void )
7   {
8      float num1, num2 ;
9
10     cout << "Type in 2 numbers. Press Enter after each number." << endl ;
11     cin >> num1 >> num2 ;
12
13     float sum ;  ◄─────────── A variable can be defined just before it is used.
14     sum = num1 + num2 ;
15     cout << num1 << " + " << num2 << " = " << sum << endl ;
16     return 0 ;
17  }
```

A sample run of this program is:

```
Type in 2 numbers. Press Enter after each number.
1.1
2.2
1.1 + 2.2 = 3.3
```

Line 11 reads the two numbers from the keyboard and line 14 assigns the result of their addition to the variable sum defined on line 13.

As shown in line 13, a variable can be defined at any point in a program, provided it is defined before it is used.

在程序的任何地方都可以定义变量，只要在使用之前定义即可。

3.2 Manipulators（流操纵符）

Manipulators are used to modify input and output data streams. The manipulator endl was used in previous programs to skip to the start of a new line on the screen.

A list commonly used manipulators is given in appendix E, but this section will look at just five of them: endl, setw, setfill, fixed and setprecision.

A manipulator can appear anywhere in a series of insertion or extraction operations. For example,

流操纵符用于对输入和输出的数据流进行修改。例如前面程序中使用的流操纵符endl，它的作用是将光标移到一个新行的起始位置。

流操纵符可以出现在一组连续的流插入和流提取运算符中的任何位置。

```
cout << endl << endl << "endl can be used anywhere" << endl ;
```

This example will skip to the start of a new line, skip another line, display the text in the double quotes and then skip to the start of the following line.

The manipulator setw is used to set the width of a data field. The width of a data field is the number of columns that the data item occupies on the screen.

这个例子是先将光标移到一个新行的起始位置，然后再换到下一行，输出双引号中的提示信息，最后再将光标移到一个新行的起始位置。

流操纵符setw用于设置数据域的宽度。数据域的宽度是指数据项在屏幕上所占的列数。

Program Example P3C

```
1   // Program Example P3C
2   // Demonstration of the setw manipulator.
3   #include <iostream>
4   #include <iomanip>
5   using namespace std ;
6
7   int main( void )
8   {
9     int num1 = 123, num2 = 4567 ;
10
11    cout << "Without setw:" << endl ;
12    cout << num1 << num2 << endl ;
13    cout << "With setw:" << endl ;
14    cout << setw( 4 ) << num1 << setw( 7 ) << num2 << endl ;
15    return 0 ;
16  }
```

The output from this program is:

```
Without setw:
1234567
With setw:
 123    4567
```

Line 4 is required for any manipulator, like `setw`, that has a value in parentheses. Other manipulators, such as `endl`, do not require this line. Without using `setw`, the two numbers are displayed beside each other without any intervening space, making it difficult to see where one number ends and the next number starts.

使用像setw这样的在圆括号中有一个参数的流操纵符时，第4行是必需的。而使用其他流操纵符（如endl）则不需要这一行。

The field width (number of columns) for `num1` is set to 4 on line 14 by inserting the manipulator `setw( 4 )`. Since `num1` is only a threedigit number and the field width is set to four, a space precedes the three digits of `num1`. Similarly the field width for `num2` is set to 7, resulting in three spaces preceding its four-digit value.

不使用setw时，这两个数就会紧挨在一起显示，中间没有任何空格，这样就很难区分第一个数是在哪里结束的，第二个数是在哪里开始的。

If the field width is set too small to display a value, the width is automatically expanded so that all the digits in the valuc are displayed. The manipulator `setfill` is used to change the "padding" character from a space to any other character. This is demonstrated in the next program.

如果域宽设置得太小不足以显示一个数值，那么系统会自动调整域宽，使其能够显示数值的所有位的数字。
流操纵符setfill用于把占位符从空格变为其他字符。

Program Example P3D

```
1   // Program Example P3D
2   // Demonstration of the setfill manipulator.
3   #include <iostream>
4   #include <iomanip>
5   using namespace std;
6
7   int main( void )
8   {
9     double num = 123.456;
10
11    cout << setw( 9 ) << setfill( '*' ) << num << endl;
12    cout << setw( 9 ) << setfill( '0' ) << num << endl;
13    cout << setw( 10 ) << num << endl;
14    return 0 ;
15  }
```

The output from this program is:

```
**123.456
00123.456
000123.456
```

Unlike `setw`, which applies only to the next data item in the output stream, the `setfill` manipulator remains in effect for all subsequent data items sent to the output stream. This is shown in line 13 where the fill character remains `'0'`, as set in line 12.

流操纵符setw只对输出流中的下一个数据项起作用，而setfill将对送入输出流中的所有后继数据项都起作用。

The manipulator `setprecision` is used to specify the number of digits of a number to display. There are two ways of using this manipulator as shown in the next program.

流操纵符setprecision用于指定要显示的数据的位数。

Program Example P3E

```
1   // Program Example P3E
2   // Program to demonstrate the setprecision and fixed manipulators.
3   #include <iostream>
4   #include <iomanip>
5   using namespace std ;
6
7   int main( void )
8   {
9     double num = 123.45678 ;
10
11    cout << num << endl ;
12    cout << setprecision( 7 ) << num << endl ;
13    cout << fixed << setprecision( 2 ) << num << endl ;
14    return 0 ;
15  }
```

The output from this program is:

```
123.457
123.4568
123.46
```

By default, the maximum number of digits displayed for a number is six, which includes digits before and after the decimal point. On line 11, num is rounded up so that there is a total of six digits displayed.

In line 12, num is rounded up so that there is a total of seven digits displayed.

In line 13, the manipulator fixed precedes the setprecision specification. In this case setprecision refers to the number of digits after the decimal point. Line 13 therefore displays the value of num rounded up to two places of decimals.

Both manipulators fixed and setprecision remain in effect for subsequent insertions into the output stream.

3.3　Single-character input and output
（单个字符的输入和输出）

Unlike pressing keys such as A, B or C on the keyboard, pressing keys such as Tab, Enter and the space bar do not display anything on the screen. These keys generate an invisible blank or white space on the screen and are consequently called whitespace characters. (Of course it is only when the background is white that white spaces are displayed. If the background is black then black spaces are generated. Nevertheless, they are still called whitespace characters.)

The following code segment inputs a character from the keyboard to the variable ch, ignoring whitespace characters.

默认情况下，数据显示的最大位数（包括小数点之前和小数点之后）是6。执行第11行时，系统对num进行了舍入处理，以便只显示6位数字。

在第12行，对num进行了舍入处理，只显示7位数字。

在第13行，流操纵符fixed放置在setprecision说明之前。这时，setprecision指定的是小数点后面的显示位数。因此，第13行显示的num的值只保留了两位小数。流操纵符fixed和setprecision将一直对输出流中的后继数据项起作用。

与在键盘上按下A、B、C这样的键不同，按下Tab、Enter和空格键不会在屏幕上显示任何字符。因此，这些键在屏幕上产生的显示结果是不可见的空白或空格，称为空白字符（当然，仅当背景是白色时，才显示白色的空格。而当背景是黑色时，产生的是黑色的空格。尽管如此，它们仍然被称为空白字符）。

```
char ch ;
cin >> ch ; // Reads the next character,
            // whitespace characters are ignored.
```

Sometimes it is required to read a single character from the keyboard regardless of whether it is a whitespace character or not. This can be done by using the manipulator noskipws (no skip whitespace).

有时, 需要从键盘读入单个字符, 不管它是否为空白字符。这时就要使用流操纵符noskipws（即no skip whitespace）来实现。

```
cin >> noskipws >> ch ; // Reads the next character,
                        // a whitespace character
                        // may be read.
```

Alternatively, the function get() associated with the input stream object cin can be used.

另外的方法是将函数get( )与输入流对象cin联合使用。

```
cin.get( ch ) ; // Also reads the next character,
                // a whitespace character may be read.
```

A function is a block of program code that carries out a specific task. Functions that are associated with an object are called *member functions* of the object.

函数是执行特定功能的程序模块。与一个对象关联的函数称为对象的成员函数。

In this case the function is pre-written and is available for use by a programmer. Chapters 7 and 8 show how to write functions and member functions.

In addition to displaying a character using >>, the output stream object cout has a member function put() that can be used to display a character.

除>>外, 还可以利用输出流对象cout的成员函数put( )来显示一个字符。

```
cout.put( ch ) ; // Display the character ch.
```

Programming pitfalls

1. Do not mix up the insertion operator << and the extraction operator >>. The insertion operator is used to insert data into the output stream; the extraction operator is used to read data from the input stream.

2. Some manipulators apply only to the next data field (e.g. `setw`); others (e.g. `setprecision`) stay in effect for all subsequent data fields.

3. The line

   ```
   #include <iomanip>
   ```

   is required for a manipulator that has a value in parentheses, e.g. `setw( 4 )`. Other manipulators, such as `endl`, do not require this line.

1. 不要混淆流插入运算符（<<）和流提取运算符（>>）。流插入运算符用于向输出流中插入数据，而流提取运算符用于从输入流中读取数据。

2. 某些流操纵符只作用于下一个数据域（例如setw）；而另外一些流操纵符（例如setprecision）对所有的后继数据域都起作用。

3. 在使用有一个参数的流操纵符时，例如setw(4)，一定要加上 "include<iomanip>" 这行代码。对于其他流操纵符（例如endl），则不需要加入这一行。

Quick syntax reference

	Syntax	Examples

Input data from the keyboard	`cin >> variable1`
	`    >> variable2`
	`... >> variablen;`
		`int num1, num2; float num3;`
		
		`cin >> num1 >> num2 >> num3;`

Input a single character from the keyboard	`cin.get( variable );`	`char char_in;` `cin.get( char_in );`
Output a single character to the screen	`cout.put( variable );`	`char char_out;` `cout.put( char_out );`
Set the width of a field	`setw( integer )`	`float num;` `cout << setw( 5 ) << num;`

Set the number of decimal places	`fixed << setprecision( integer )`	`cout << fixed`
		`<< setprecision( 2 )`
		`<< num ;`

Set a fill character	`setfill( character )`	`cout << setw( 5 )`
		`<< setfill('0')`
		`<< num ;`

Exercises

1. Write a program to input four numbers and display them in reverse order.

2. Write a program that inputs a number of hours and displays the equivalent number of weeks, days and hours. For example, an input of 553 should display 3 weeks, 2 days and 1 hour.

3. Assuming the human heart rate is seventy-five beats per minute, write a program to ask a user their age in years and to calculate the number of beats their heart has made so far in their life. Ignore leap years.（leap years：闰年）

4. Write a program to accept a temperature in degrees Fahrenheit and convert it to degrees Celsius. Your program should display the following prompt:

```
Enter a temperature in degrees Fahrenheit:
```

You will then enter a decimal number followed by the Enter key.

The program will then convert the temperature by using the formula

Celsius = (Fahrenheit − 32.0) * (5.0 / 9.0)

Your program should then display the temperature in degrees Celsius using an appropriate message.

5. Make changes to the program in exercise 4 to accept the temperature in degrees Celsius and convert it to degrees Fahrenheit.

（Fahrenheit：华氏温度；Celsius：摄氏温度）

6. Write a program to accept a distance in kilometres and display the equivalent distance in miles.

(1 mile = 1.609 344 kilometres.)

7. Write a program to input three floating-point numbers from the keyboard and to calculate

(a) their sum and

(b) their average.

Display the results to three decimal places.

8. Write a program to input a year and display the century that year is in.

For example, if the input is 2016 then the output should be 21.

9. Write a program to read in two numbers from the keyboard and to display the result of dividing the second number into the first.

For example, if the input is 123 and 12, the result should be displayed in the following format:

```
123 divided by 12 = 10 Remainder = 3
```

(Hint: use the modulus operator % to get the remainder 3, and use integer division to get the quotient 123.)（modulus operator：求余运算符）

Chapter Four
Selection and Iteration
第 4 章　选择与循环

All the programs written so far execute one statement after the other, starting at the first statement and finishing at the last. Only the simplest of problems can be solved using this sequential top-to-bottom approach. Selection and iteration program constructs are used to modify this sequential execution of program statements.

到目前为止，我们编写的所有程序都是一条语句接着一条语句顺序执行的，从第一条语句开始，一直执行到最后一条语句结束。使用这种自顶向下的顺序结构只能解决最简单的问题，要解决复杂的问题必须使用能够改变程序语句执行顺序的选择和循环结构。

4.1　Selection（选择）

4.1.1　The `if` statement

An `if` statement starts with the keyword `if` followed by an expression in parentheses. If the expression is found to be true, then the statement following the `if` is executed. If the expression is untrue, then the statement following the `if` is not executed. For example:

if语句以关键字if作为开始，后跟一个用圆括号括起来的表达式。如果表达式为真，那么执行if后面的语句。如果表达式为假，那么不执行if后面的语句。

```
if ( account_balance < 0 )
  cout << "Your account is in the red" << endl ;
```

This statement tests whether the value of `account_balance` is less than 0 or not. If the value is less than 0, then the message is displayed; otherwise the message is not displayed.

The < is called a *relational operator*. The full list of relational operators is given in Table 4.1.

这里的＜称为关系运算符。表4.1列出了全部的关系运算符。

Table 4.1　List of Relational Operators

Operator	Meaning
==	equivalent to
!=	not equal to
<	less than
>	greater than
<=	less than or equal to
>=	greater than or equal to

The next program demonstrates the use of if statements with < and >=.

Program Example P4A

```
1  // Program Example P4A
2  // Program to demonstrate if statements.
3  #include <iostream>
4  using namespace std ;
5
```

```
6   int main( void )
7   {
8     float account_balance ;
9
10    cout << "What is your account balance? " ;
11    cin >> account_balance ;
12
13    if ( account_balance < 0 )
14    cout << "Your account is in the red" << endl ;
15    if ( account_balance >= 0 )
16    cout << "Your account is in the black" << endl ;
17    return 0 ;
18  }
```

When this program is run, it will ask the user to enter an account balance. If the account balance is negative, the statement on line 14 is executed; if it is greater than or equal to 0, then the statement on line 16 is executed. Here is a sample run of this program:

```
What is your account balance? -100
Your account is in the red
```

4.1.2 The `if-else` statement

With the simple `if` statement there is a choice of either executing a statement or skipping it. With an `if-else` there is a choice of executing one or other of two statements.

使用简单的if语句，需要做出的选择是：要么执行一条语句，要么跳过它。使用if-else语句，需要做出的选择是：在两条语句中选择其中的一条来执行。

Program Example P4B
```
1   // Program Example P4B
2   // Program to demonstrate the use of if-else.
3   #include <iostream>
4   using namespace std ;
5
6   int main( void )
7   {
8     float account_balance ;
9
10    cout << "What is your account balance? " ;
11    cin >> account_balance ;
12
13    if ( account_balance < 0 )
14    cout << "Your account is in the red" << endl ;
15    else
16    cout << "Your account is in the black" << endl ;
17    return 0 ;
18  }
```

In this program, if the value of `account_balance` is less than 0, line 14 is executed; otherwise the line 16 is executed.

4.1.3　Compound statements

A *compound statement* is one or more statements enclosed in braces { and }. A compound statement can be used anywhere a single statement can be used.

用一对花括号括起来的一条或多条语句，称为复合语句。

Program Example P4C

```cpp
1  // Program Example P4C
2  // Program to demonstrate the formation and use of a
3  // compound statement.
4  #include <iostream>
5  #include <iomanip>
6  using namespace std ;
7
8  int main( void )
9  {
10   float account_balance, interest ;
11   const float overdraft_rate = 10.0 ;
12
13   cout << "What is your account balance? " ;
14   cin >> account_balance ;
15
16   if ( account_balance < 0 )
17   {
18     cout << "Your account is in the red" << endl ;
19     interest = -account_balance * overdraft_rate / 100.0 ;
20     cout << "The interest charged is "
21          << fixed << setprecision( 2 ) << interest << endl ;
22   }
23   else
24   {
25     cout << "Your account is in the black" << endl ;
26     cout << "There is no interest charged" << endl ;
27   }
28   return 0 ;
29 }
```

The statements within a compound statement are usually indented for appearance purposes. For example, lines 18 to 20 are two spaces to the right of the opening brace on line 17.

This program will ask for a bank balance and will then calculate an overdraft charge, assuming an overdraft rate of 10 per cent. The balance is tested in line 16. If the balance is less than 0, the statements on lines 18 to 21 are executed. If the balance is not less than 0, then the statements on lines 25 and 26 are executed.

Here is a sample run of this program:

为了使程序层次清晰，通常将复合语句中的语句向右缩进。

```
What is your account balance? -100
Your account is in the red
The overdraft charge is 10.00
```

4.1.4 Logical operators

There are three logical operators (see Table 4.2) for use in `if` statements:

在if语句中可以使用三种逻辑运算符：&&（与），‖（或），!（非）。

Table 4.2   Logical Operators

Logical operator	Meaning
&&	AND
‖	OR
!	NOT

The logical operators `&&` (AND) and `‖` (OR) are used to combine tests within an `if` statement.

`&&` is used to join two simple conditions together; the resulting compound condition is only true when *both* simple conditions are true. If `‖` is used to join two simple conditions, the result is true if *either* or *both* are true.

The logical NOT operator `!` is used to reverse the result of an `if` statement. If the result is true, then it becomes false, and if it is false, then it becomes true.

逻辑运算符&&和‖用于在if语句中实现组合测试。
&&用于将两个简单的条件组合，得到的复合条件是，当且仅当两个简单条件均为真时才为真。而‖将两个简单的条件组合得到的复合条件是，只要两者之一或两者均为真就为真。
逻辑非运算符!用于将一个结果取反。如果结果为真，那么就将其变为假，否则将其变为真。

Examples:

```
if ( a == 0 && b == 0 )    // Example of &&
  cout << "both a AND b are zero" ;

if ( a == 0 || b == 0 )    // Example of ||
  cout << "a OR b is zero" ;

if ( ! ( a == 0 ) )        // Example of !
  cout << "a is not zero" ;
```

4.1.5 Nested `if` statements

When an `if` statement occurs within another `if` statement it is called a *nested* if statement. For example:

当一条if语句出现在另一条if语句中间时，称为嵌套的if语句。

```
if ( a == 0 && b == 0 )
  cout << "Both a AND b are zero" ;
```

can be rewritten using a nested `if` as follows:

```
if ( a == 0 )
  if ( b== 0 )
    cout << "Both a AND b are zero" ;
```

4.1.6 The `switch` statement

The `switch` statement provides an alternative to a series of `if-else` statements, which can become quite complex to follow.

The next program emulates a four-function calculator. For each calculation the user inputs two numbers and an operator. For example, if the input is 5 + 3 then the program will display 8.

switch语句是一个if-else语句序列的另一种表示形式，但是比if-else语句序列要简单得多。

Program Example P4D

```
1   // Program Example P4D
2   // Simple four-function calculator.
3   // This program illustrates the use of the switch statement.
4   #include <iostream>
5   using namespace std ;
6
7   int main( void )
8   {
9     char op ;
10    float num1, num2, answer ;
11
12    cout << "Please enter an arithmetic expression (e.g. 1 + 2) " ;
13    cin >> num1 >> op >> num2 ;
14
15    switch( op )
16    {
17      case '+' :
18        answer = num1 + num2 ;
19        cout << num1 << " plus " << num2 << " equals "
20             << answer << endl ;
21        break ;
22
23      case '-' :
24        answer = num1 - num2 ;
25        cout << num1 << " minus " << num2 << " equals "
26             << answer << endl ;
27        break ;
28
29      case '*' :
30        answer = num1 * num2 ;
31        cout << num1 << " multiplied by " << num2 << " equals "
32             << answer << endl ;
33        break ;
34
35      case '/' :
36        answer = num1 / num2 ;
37        cout << num1 << " divided by " << num2 << " equals "
38             << answer << endl ;
39        break ;
40
41      default :
42        cout << "Invalid operator" << endl ;
43    }
44    return 0 ;
45  }
```

A sample run of this program will produce the following output:

```
Please enter an arithmetic expression (e.g. 1+2) 5+3
5 plus 3 equals 8
```

The `switch` statement is equivalent to a series of if-else statements. The variable or expression to be tested is placed in parentheses after the keyword `switch`. Unfortunately, this variable or expression can only be of type `char` or `int`, which somewhat limits the usefulness of `switch`.

As many cases as are required are then enclosed within braces. Each case begins with the keyword `case`, followed by the value of the variable and a colon. The value of the variable is compared with each case value in turn. If a match is found, then the statements following the matching `case` are executed.

Once a match is found and the appropriate statements are executed, the `break` statement terminates the `switch` statement. Without the `break` statement, execution would continue to the end of the `switch` statement. The `break` can be omitted when the same statements are to be executed for several different cases. For example, in the last program either *, x or X can be used to indicate multiplication by modifying the `switch` statement as follows:

```
case '*':
case 'x':
case 'X':
  answer = num1 * num2;
  cout << num1 << " multiplied by " << num2 << " equals"
      << answer << endl;
  break;
```

Either *, x or X will execute the same statements.

If no case matches the value of the `switch` variable, the `default` case is executed. In this program the `default` case is used to trap an invalid operator.

4.1.7　The conditional operator ?:

The conditional operator `?:` is a short form of `if-else`.

The following program reads two values from the keyboard and finds the larger of the two using the `?:` operator.

switch语句等价于一个if-else语句序列。待测试的变量或者表达式被置于关键字switch后面的圆括号内。但遗憾的是，这个变量或者表达式的类型只能是char或int，这在某种程度上限制了switch的应用。

花括号内包含很多种情况，每种情况都从关键字case开始，其后是一个值和一个冒号。变量的值依次与每个case后的值进行比较，如果发现与某个case后的值相匹配，就执行其后的语句。

一旦找到一个匹配，就执行其后的语句序列，直到遇到break语句为止。如果没有break语句，那么将一直执行到switch语句的末尾。当几种不同的情况需要执行相同的语句时，可以省略break语句。例如，对于程序P4D，当*、x或X均可用于表示乘法符号时，就需要对程序中的switch语句进行如下修改。

如果找不到与switch变量的值相匹配的情况，就执行default后面的语句。在这个程序中，default情况用于处理无效的运算符。

条件运算符?:是if-else的简写格式。

Program Example P4E

```
1   // Program Example P4E
2   // Demonstration of the conditional operator ?:
3   #include <iostream>
4   using namespace std ;
5
6   int main( void )
7   {
8   float max, num1, num2 ;
9
10     cout << "Type in two numbers. Press Enter after each number."
11         << endl ;
```

```
12   cin >> num1 >> num2 ;
13
14   // Assign max to the larger of the two numbers.
15   max = ( num1 > num2 ) ? num1 : num2 ;
16   cout << "The larger number is " << max << endl ;
17   return 0 ;
18 }
```

Line 15 is just a shorthand way of writing

```
if ( num1 > num2 )
  max = num1 ;
else
  max = num2 ;
```

A sample run of this program follows:

```
Type in two numbers. Press Enter after each number.
1
2
The larger number is 2
```

4.2   Iteration（循环）

Iterative control statements allow you to execute one or more program statements repeatedly. C++ has three iterative control statements: the while, the do-while and the for statements.

循环控制语句可以使程序重复地执行一条或多条语句。C++中有三种循环控制语句：while语句、do-while语句和for语句。

4.2.1   The while statement

The while statement causes one or more statements to repeat as long as a specified expression remains true. The next program demonstrates the while statement by inputting a series of numbers and displaying a running total of the numbers. The program stops when a 0 is input.

只要while语句中的特定表达式的值保持为真，一条或多条语句就将被反复执行。下面的程序演示了while语句的功能。程序请用户输入一组数据，然后显示每次执行累加运算的结果。当用户输入0时，程序结束。

Program Example P4F
_____

```
1   // Program Example P4F
2   // Program to demonstrate the use of the while statement.
3   // This program reads in a series of numbers from the
4   // keyboard, prints a running total, and stops when a 0
5   // is entered.
6   #include <iostream>
7   using namespace std ;
8
9   int main( void )
10  {
11     float num, total ;
12
13     total = 0 ;
14     num = 1 ;
15
16     while( num != 0 )
```

As long as num *is not 0, this condition is true and the loop continues.*

```
17      {
18          cout << "Please enter a number(0 to end) "
19          cin >> num ;
20          total += num ;
21          cout << "The running total is " << total << endl << endl ;
22      }
23
24      cout << "The final total is " << total << endl ;
25      return 0 ;
26  }
```

As long as the condition is true, the statements between { and } are executed.

The statements enclosed within the braces { and } are executed repeatedly while the control expression n != 0 is true. The repeated execution of one or more program statements is called a *program loop*. The braces forming the loop may be omitted if there is only one statement in the loop.

当控制表达式n != 0为真时，花括号内的语句将被重复执行。重复执行一条或者多条语句称为程序循环。如果循环体内仅有一条语句，那么可以省略循环体的花括号。

A sample run of this program displays the following:

```
Please enter a number 12
The running total is 12

Please enter a number 6.4
The running total is 18.4

Please enter a number −1.25
The running total is 17.15

Please enter a number 0
The running total is 17.15

The final total is 17.15
```

The statements on lines 18 to 21 are executed repeatedly while the value of the variable num is not 0. When num becomes 0, the loop stops and the statement on line 24 is executed.

The control expression in a while loop is tested before the statements in the loop are executed. The sequence in a while loop is as follows:

1. Evaluate the control expression.
2. If the control expression is true, execute the statements in the loop and go back to 1.
3. If the control expression is false, exit the loop and execute the next statement after the loop.

It is important to note that if the first evaluation of the control expression is false, the statements in the loop are never executed. This is the purpose of giving the variable num a non-zero value in line 14. Line 14 places a value of 1 into num, but any non-zero value would do.

while循环中控制表达式的值是在执行循环体内的语句之前测试的。while循环的执行过程如下：

1. 计算控制表达式的值。
2. 如果控制表达式的值为真，那么就执行循环体中的语句，然后返回步骤1。
3. 如果控制表达式的值为假，就退出循环，执行循环体后面的语句。

值得注意的是，如果第一次计算的控制表达式的值为假，那么循环体内的语句将永远都不会被执行。这就是在第14行给变量num赋值为非零值的目的所在。第14行给变量赋值的值是1，但其实给它赋予其他任意的非零值都是可以的。

4.2.2 The `do-while` loop

In a `while` loop the control expression is tested *before* the statements in the loop are executed. The test in a `do-while` loop is done *after* the statements in the loop are executed. This means that the statements in a `do-while` loop are executed at least once. The sequence in a `do-while` loop is as follows:

1. Execute the statements in the loop.
2. Evaluate the control expression.
3. If the control expression is true then go back to 1.
4. If the control expression is false then exit the loop and execute the next statement after the loop.

The next program replaces the `while` loop in program P4F with a `do-while` loop.

在 while 循环中，控制表达式的值是在执行循环体内的语句之前测试的。而 do-while 循环中的控制表达式的值则是在执行循环体内的语句之后测试的，这就意味着 do-while 循环中的语句至少会被执行一次。do-while 循环的执行过程如下：

1. 执行循环体中的语句。
2. 计算控制表达式的值。
3. 如果控制表达式的值为真，那么返回步骤 1。
4. 如果控制表达式的值为假，则退出循环，执行循环体后面的语句。

Program Example P4G

```cpp
1  // Program Example P4G
2  // Program to demonstrate a do-while loop.
3  // This program will read in a series of numbers from the
4  // keyboard, prints a running total, and stops when a 0
5  // is entered.
6  #include <iostream>
7  using namespace std ;
8
9  int main( void )
10 {
11   float num, total ;
12
13   total = 0 ;
14
15   do
16   {
17     cout << "Please enter a number " ;
18     cin >> num ;
19     total += num ;
20     cout << "The running total is " << total << endl << endl ;
21   }
22   while ( num != 0 ) ;
23
24   cout << "The final total is " << total << endl ;
25   return 0 ;
26 }
```

The statements between { and } are executed at least once.

Using a `do-while` loop means there is no need to initialise the variable num, because the loop is executed at least once. The output from this program is the same as for program P4F.

因为循环体至少被执行一次，所以使用 do-while 循环时无须对变量 num 进行初始化。

4.2.3 The `for` statement

The `for` statement is used to execute one or more statements a specified number of times. The general format of the `for` statement is:

for 语句用于将一条或者多条语句执行指定的次数。

```
for ( initial expression; continue condition; increment expression )
{
  ...
  // one or more statements.
  ...
}
```

The `for` statement consists of three expressions enclosed in parentheses and separated by semicolons.

The `initial expression` is executed once at the beginning of the loop. The loop continues while the `continue expression` is true (a non-zero value) and terminates when `continue expression` becomes false (a zero value). This test occurs before each pass, including the first pass, through the loop. The `increment expression` is executed at the end of every pass through the loop.

The braces { and } are needed only when there is more than one statement in the loop.

The next program displays a table of squares and cubes from 1 to 5.

for语句由圆括号内以分号分隔的三个表达式组成。

初始化表达式仅在循环开始时执行一次。当循环继续条件表达式的值为真（一个非零值）时，继续执行循环体，当循环继续条件表达式的值变为假（一个零值）后，结束循环体的执行。在每次（包括第一次）执行循环体之前，都要对循环继续条件测试一次。每次循环体执行完以后，都要执行一次增值表达式。

仅在循环体内有多条语句时，花括号{和}才是必需的。

Program Example P4H

```
1   // Program Example P4H
2   // Program to display a table of squares and cubes.
3   #include <iostream>
4   #include <iomanip>
5   using namespace std ;
6
7   int main( void )
8   {
9     cout << "Number Square Cube" << endl ;
10    cout << "----------------------" << endl ;
11
12    for ( int i = 1 ; i < 6 ; i++ )
13    {
14      cout << setw( 3 ) << i
15           << setw( 10 ) << i * i
16           << setw( 8 ) << i * i * i << endl ;
17    }
18    return 0 ;
19  }
```

The output from this program is:

```
Number    Square   Cube
-----------------------
   1         1       1
   2         4       8
   3         9       27
   4        16       64
   5        25       125
```

The statements within the braces on lines 13 and 17 are executed five times. The loop contains only one statement, so the braces may be omitted. However, the braces are useful in that they clearly show the body of the loop.

The variable i is defined within the parentheses on line 12, but it could also be defined before the for statement as in the following:

```
int i ;
for ( i = 1 ; i < 6 ; i++ )
...
```

The for statement on line 12 causes the loop to be executed 5 times, with the variable i starting at 1 and continuing while the value of i is less than 6. Each time the loop is completed the value of i is incremented by 1. When the value of i becomes 6, the loop terminates. There are many variations you can add to the simple for on line 12. Try modifying the program in each of the following ways:

1. To display the table for numbers from 10 down to 1, change line 12 to:

```
for ( int i = 10 ; i > 0 ; i -- )
```

Here i starts at 10 and is decremented at the end of each pass through the loop until it eventually becomes 0.

2. To display the table for even numbers from 2 to 10, change line 12 to:

```
for ( int i = 2 ; i <= 10 ; i += 2 )
```

In this for statement the variable i is initialised to 2 and increases by 2 each time through the loop, giving i the values 2 4 6 8 10. Either initial expression or increment expression, or both, may consist of multiple expressions separated by commas. For example:

```
for ( int i = 0, int j = 0 ; i < 10 ; i++, j++ )
```

This loop initialises both i and j to 0 and increments both of them at the end of each pass through the loop.

第13行和第17行的花括号内的语句被执行了5次。由于循环体内仅有一条语句，因此，花括号可以省略。但是使用花括号有助于清晰地表示循环体。

变量i是在第12行的圆括号内定义的，但也可以在for语句之前定义。

由于循环从变量i初始化为1开始，在i值小于6之前，将持续循环，因此第12行的for语句使循环体被执行了5次。每执行完一次循环体，都要将变量i的值增1。当变量i的值变成6时，循环结束。

Any or all of the three expressions may be omitted from a `for` statement, but the two semicolons must always be present in the statement. For example, the statement `for ( ; ; )` will create an infinite loop because there is no condition to end the loop.

4.2.4　Nested loops

A `for` loop can contain any valid statements, including another `for` loop. When a loop is contained within another loop it is called a *nested loop*.

The next program displays a 12 by 12 multiplication table using a nested loop.

for语句中的任何一个或者所有这三个表达式都可以省略，但是分号不能省略。例如，语句 for(;;)，由于没有可以使循环结束的条件，所以它将使循环成为一个无限循环（即死循环）。

一个for循环可以包含任何有效的语句，包括包含另一个for循环。当一个循环包含在另一个循环中时，这种形式称为嵌套循环。

Program Example P4I

```cpp
1   // Program Example P4I
2   // Program displays a 12 x 12 multiplication table
3   // using a nested loop.
4   #include <iostream>
5   #include <iomanip>
6   using namespace std ;
7
8   int main( void )
9   {
10    int i, j ;
11
12    cout << " " ;
13
14    for ( i = 1 ; i <= 12 ; i++ )
15    {
16      cout << setw( 5 ) << i ;
17    }
18
19    cout << endl << " +" ;
20
21    for ( i = 0 ; i <= 60 ; i++ )
22    {
23      cout << '-' ;
24    }
25
26    for ( i = 1 ; i <= 12 ; i++ )       // Start of outer loop.  <--+
27    {                                   //                           |
28      cout << endl << setw( 2 ) << i << " |" ;   //                  |
29      for ( j = 1; j <= 12 ; j++ )      // Start of inner loop.<-+   |
30      {                                 //                       |   |
31        cout << setw( 5 ) << i * j ;    //                       |   |
32      }                                 // End of inner loop. <--+   |
33    }                                   // End of outer loop.    <--+
34
35    cout << endl ;
36    return 0 ;
37  }
```

This program will display the following table:

	1	2	3	4	5	6	7	8	9	10	11	12
+	---	---	---	---	---	---	---	---	---	----	----	----
1 |	1	2	3	4	5	6	7	8	9	10	11	12
2 |	2	4	6	8	10	12	14	16	18	20	22	24
3 |	3	6	9	12	15	18	21	24	27	30	33	36
4 |	4	8	12	16	20	24	28	32	36	40	44	48
5 |	5	10	15	20	25	30	35	40	45	50	55	60
6 |	6	12	18	24	30	36	42	48	54	60	66	72
7 |	7	14	21	28	35	42	49	56	63	70	77	84
8 |	8	16	24	32	40	48	56	64	72	80	88	96
9 |	9	18	27	36	45	54	63	72	81	90	99	108
10 |	10	20	30	40	50	60	70	80	90	100	110	120
11 |	11	22	33	44	55	66	77	88	99	110	121	132
12 |	12	24	36	48	60	72	84	96	108	120	132	144

The loop in lines 14 to 17 displays the numbers 1 to 12 across the screen, and the loop in lines 21 to 24 displays the hyphens beneath them. The remainder of the program uses a nested loop to display the numbers in the table.

The outer loop (lines 26 to 33) starts with i at 1. Line 28 displays a 1 and the vertical stroke character | at the left of the screen.

The inner loop (lines 29 to 32) is then executed to completion, with j starting at 1 and ending when j exceeds 12. Each iteration of the inner loop displays a number in the multiplication table.

When the inner loop is completed, the outer loop regains control, and i is incremented to 2.

Line 28 then displays 2 | at the left of the screen, and the inner loop on lines 29 to 32 is executed again.

The program continues until the outer loop is completed when the value of i exceeds 12.

第14行至第17行的循环在屏幕的水平方向上显示数字1至12。第21行至第24行的循环在这些数字的下面显示连字符-。程序的其余部分使用一个嵌套循环显示表格中的数字。

外层循环（第26行至第33行）从i等于1开始，第28行在屏幕的左端显示一个1和一个竖线|。

接下来，执行内层循环（第29行至第32行），内层循环从j等于1开始，到j大于12时结束。每执行一次内层循环，就显示乘法表中的一个数。

当内层循环执行完毕以后，又转向外层循环控制，i的值增加到2。第28行在屏幕的左端显示2|，然后，又开始执行第29行至第32行的内层循环。

程序继续执行，直到i的值超过12时，外层循环结束。

Programming pitfalls

1. There is no ; immediately after an `if` statement. For example:

```
if ( account_balance < 0 ) ; // Misplaced semicolon.
  cout << "Your account is in the red" << endl ;
```

should be:

```
if ( account_balance < 0 )
  cout << "Your account is in the red" << endl ;
```

In the first case the message `Your account is in the red` is always displayed, regardless of the value in `account_balance`. The reason for this is that the `cout` statement is not controlled by the `if`. The `if` only controls the empty statement mistakenly made by the misplaced semicolon.

2. There is no ; after `switch`.
3. When testing for equality use ==, not =.
4. Each `else` is matched with the previous `if`.
5. For each opening brace { there will be a closing brace }.
6. Braces are necessary to control the execution of a set of statements with an `if` statement.

   For example:

```
if ( a == b )
  a = 1 ;
  b = 2 ;
```

In this example, the statement `a = 1` is executed only if `a` and `b` are equal. However, the statement `b = 2` is always executed, regardless of the values of `a` and `b`. To execute both statements when `a` and `b` are equal the braces are required:

```
if ( a == b )
{
  a = 1 ;
  b = 2 ;
}
```

7. The logical operators (`&&` and `||`) evaluate the smallest number of operands needed to determine the result of an expression. This means that some operands of the expression may not be evaluated. For example, in

```
if ( a > 1 && b++ > 2 )
```

the second operand, `b++`, is evaluated only if the condition `a > 1` is logically true.

1. 紧随if语句之后不应有分号。

2. 紧随switch语句之后不应有分号。
3. 在测试两个值是否相等时，应使用==，而不能使用=。
4. 每个else都与其前面最邻近的那个if语句配对。
5. 对于每个左花括号{，必定有一个与其配对的右花括号}。
6. 要想在if语句中控制一个语句序列的执行，必须使用花括号。

7. 在计算含有逻辑运算符（&&和‖）的表达式时，尽量使其能够根据最少的操作数确定整个表达式的值，这意味着表达式中的某些操作数的值就无须再进行计算了。

8. There is no semicolon immediately after the `while` or `for` statements. For example:

```
for ( i = 0; i < 10; i++ ) ; // Misplaced semicolon.
  cout << "The value of i is " << i ;
```

This loop does not contain any statements and will not display the values 0 to 9, as expected. Only the final value of `i` (=10) will be displayed. The reason for this is that a semicolon immediately after a `while` or `for` statement makes the body of the loop empty, i.e. containing no executable statements.

9. Be careful in specifying the terminating condition in a `for` loop. For example:

```
for ( int i = 0 ; i == 10 ; i++ )
// This loop does nothing.
  cout << "The value of i is " << i << endl ;
```

This loop does nothing, because `i == 10` is false at the start of the loop (`i` is in fact `0`) and the loop terminates immediately. Replace `i == 10` with `i < 10` or `i != 10` and the loop will execute ten times.

10. There is no semicolon after `while` in a while loop, but there is in a do-while loop.
    See line 16 of program P4F and line 22 of program P4G.

11. There is a limit to the precision with which floating-point numbers are represented. This is important when testing a floating-point number for equality in an `if` or in a `for` loop. For example, consider the following loop:

```
float f ;
for ( f = 0.0 ; f != 1.1 ; f += 0.1 )
{
  ...
  // statement(s) in the loop.
  ...
}
```

On most computers this will result in an infinite loop. The reason for this is that `f` may never equal `1.1` exactly. You can allow for this situation by writing the loop as:

```
for ( f = 0.0 ; f <= 1.1 ; f += 0.1 )
{
...
// statement(s) in the loop.
...
}
```

8. 紧随 while 或 for 语句之后不应有分号。

9. 仔细定义 for 循环的终止条件。

10. while 循环的 while 之后不应有分号，但是 do-while 循环的 while 之后有分号。

11. 浮点数所表示的精度是有限的。在 if 语句或者 for 语句中测试浮点数是否相等时，了解这一点是非常重要的。

Quick syntax reference

	Syntax	Examples

if-else	```if (condition)```
	```{```
	``` statement(s) ;```
	```}```
	```else```
	```{```
	``` statement(s) ;```
	```}```
		```if (n > 0)```
		```{```
		``` average = total / n ;```
		``` cout << average ;```
		```}```
		```else```
		``` average = 0 ;```

?:	```variable = (condition) ? v1 : v2 ;```	```max = (n1 > n2) ? n1 : n2 ;```

switch	```switch (expression)```
	```{```
	```case value1 :```
	``` statement(s) ;```
	``` break ;```
	```case value2 :```
	``` statement(s) ;```
	``` break ;```
	```default :```
	``` statement(s) ;```
	```}```
		```char traffic_light ;```
		```...```
		```switch(traffic_light)```
		```{```
		```case 'R':```
		```case 'r':```
		``` cout << "Red: STOP" ;```
		``` break ;```
		```case 'G':```
		```case 'g':```
		``` cout << "Green: GO" ;```
		``` break ;```
		```case 'A':```
		```case 'a':```
		``` cout << "Amber: READY" ;```
		``` break ;```
		```default:```
		``` cout << "FAULT" ;```
		```}```

while	```while (condition)```
	```{```
	``` statement(s) ;```
	```}```
		```// Read and total until n is 0.```
		```int n = 1 ;```
		```int total = 0 ;```
		```while (n != 0)```
		```{```
		``` cin >> n ;```
		``` total += n ;```
		```}```

do-while	```do```
	```{```
	``` statement(s) ;```
	```}```
	```while (condition) ;```
		```// Read and total until n is 0.```
		```int n ;```
		```int total = 0 ;```
		```do```
		```{```
		``` cin >> n ;```
		``` total += n ;```
		```}```
		```while (n != 0) ;```

(cont.)

	Syntax	Examples

for	for (initial expression ;
	    continue condition ;
	    increment expression)
	{
	    statement(s) ;
	}
		// Read and total 10 numbers.
		int total = 0 ;
		for (int i = 0 ; i < 10 ; i++)
		{
		  cin >> n ;
		  total += n ;
		}

Exercises

1. Rewrite the following `if-else` using a `switch` statement:

```
if ( marriage_status == 'S' )
  cout << "single" ;
else if ( marriage_status == 'M' )
  cout << "married" ;
else if ( marriage_status == 'W' )
  cout << "widowed" ;
else if ( marriage_status == 'E' )
  cout << "separated" ;
else if ( marriage_status == 'D' )
  cout << "divorced" ;
else
  cout << "error: invalid code" ;
```

2. The following program segment displays an appropriate message depending on the values of three integers: n1, n2, and n3.

```
if ( n1== n2 )
if (n1 == n3 )
cout << "n1, n2 and n3 have the same value" << endl ;
else
cout << "n1 and n2 have the same value" << endl ;
else if ( n1 == n3 )
cout << "n1 and n3 have the same value" << endl ;
else if ( n2 == n3 )
cout << "n2 and n3 have the same value" << endl ;
else
cout << "n1, n2 and n3 have different values" << endl ;
```

Use spaces to improve the readability of this code.

To test the various branches in this code you will need to construct five sets of test data, each set testing one of the branches. Construct the five sets of test data for n1, n2, and n3.

3. Write a program to read in two integers and check if the first integer is evenly divisible by the second. (Hint: use the modulus operator %.)

4. Input two numbers and find the smaller of the two using the conditional operator ?:.

5. In a triangle, the sum of any two sides must be greater than the third side.（triangle：三角形）

Write a program to input three numbers and determine if they form a valid triangle.

6. Input a person's height in centimetres and weight in kilograms and display a message indicating that they are either underweight, overweight or normal weight. As an approximation, a person is underweight if their weight is less than their height divided by 2.5 and they are overweight if their weight is greater than their height divided by 2.3.

7. Write a program that reads a single numeral from the keyboard and displays its value as a word. For example, an input of 5 will display the word 'five'.

8. Write a program to input a number 1 to 7 from the keyboard, where 1 represents Sunday, 2 Monday, 3 Tuesday, etc. Display the day of the week corresponding to the number typed by the user. If the user types a number outside the range 1 to 7, display an error message.

9. Add the increment operator (I or i) and the decrement operator (D or d) to the simple calculator program P4D.

10. Write a program to input the time of day in Ireland and display the equivalent time in Washington (−5 hours), Moscow (+ 3 hours), and Beijing (+ 8 hours). Input the time in the 24-hour format, e.g. 22:35 (11:35 p.m.).

11. Write a program to display the effects of an earthquake based on the Richter scale value:

Richter scale value	Effects
Less than 4	Little.
4.0 to 4.9	Windows shake.
5.0 to 5.9	Walls crack; poorly built buildings are damaged.
6.0 to 6.9	Chimneys tumble; ordinary buildings are damaged.
7.0 to 7.9	Underground pipes break; well-built buildings are damaged.
More than 7.9	Ground rises and falls in waves; most buildings are destroyed.

（Richter scale value：里氏震级）

12. What is the output from the following?

```
for ( int j = 1, int i = 10 ; i > 0 ; i /= 2, j++ )
  cout << i << j ;
```

13. Modify program P4F to calculate the average along with the total of the numbers entered.

14. Rewrite the following using a `for` loop.

```
int i = 0, total = 0 ;
while ( i < 10 )
{
   cin >> n ;
   total += n ;
   i++ ;
}
```

15. What is displayed when the following program is run and the number 1234 is entered?

```
int num ;
cout << "Please enter a number " ;
cin >> num ;
do
{
  cout << num % 10 ;
```

```
    num /= 10 ;
  }
while ( num != 0 ) ;
```

16. The following program segment is intended to compute 0.1 + 0.2 + 0.3 + ... + 99.8 + 99.9. It contains a flaw. What is it, and how would you correct it?

```
float sum = 0.0 ;
float i = 0.1 ;
while ( i != 100.0 )
{
  sum += i ;
  i += 0.1 ;
}
```

17. What is the output from the following?

    (a)
```
for (int i = 0 ; i < 5 ; i++ )
  for ( int j = i ; j < 5 ; j++ )
    cout << i << j << endl ;
```
    (b)
```
for ( int i = 0 ; i < 5 ; i++ )
  for ( int j = 0 ; j < 5-i ; j++ )
    cout << i << j << endl ;
```

18. Write a `for` loop to

    (a) display the numbers 0, 5, 10, 15, ..., 100.

    (b) display the numbers 1, 2, 4, 8, 16, ..., 1024.

19. Write a program to find the sum of all the odd integers in the range 1 to 99.

20. Write a program that displays all the numbers from 5 to 50 that are divisible by 3 or 5.

21. Write a program to display all the hour and minute values in a 24-hour clock, i.e. 0:00 0:01 ... 23:59. How would you display the values in fifteen-minute intervals?

22. Write a program to display the following two triangles:

```
*
**
***
****


   *
  **
 ***
****
```

    Input the size of the triangles from the keyboard.

23. Write a program that lets a teacher enter the percentage marks for a class. The teacher enters a negative mark to indicate that there are no more marks to be entered. Once all the marks have been entered, the program displays the average mark for the class.（negative mark：负数标记）

Chapter Five
Arrays and Structures
第 5 章 数组和结构体

5.1 Arrays（数组）

5.1.1 Introduction

An array is a group of variables of the same data type, such as ten `ints`, fifteen `chars` or a hundred `floats`. For example, an array of ten integers is defined by:

数组由一组相同类型的变量组成，如10个int型变量，15个char型变量，100个float型变量。

```
int numbers[10] ; // Defines an array of 10 ints called numbers.
```

The individual values or elements in the array are all held in memory under one name, i.e. the array name. The array name can be any valid variable name.

Visually the array `numbers` is stored in memory like this:

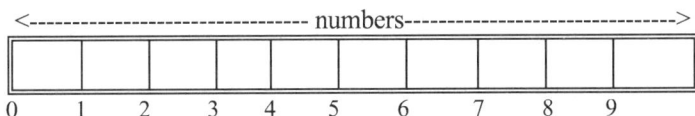

数组中的每个数值或元素都被保存在同一个名字（即数组名）之下的内存中，数组名可以是任何有效的变量名。

```
<----------------------------- numbers----------------------------->
|    |    |    |    |    |    |    |    |    |    |
 0    1    2    3    4    5    6    7    8    9
```

Each individual element of the array is accessed by reference to its position in the array relative to the first element of the array. The position of an element in an array is called the *index* or *subscript*. For example, the first element in the array `numbers` has an index value of 0, and the last element has an index value of 9. Note that the index of the array `numbers` goes from 0 to 9, not 1 to 10.

To refer to a particular element of an array, the array name and the index in brackets is used. In the array `numbers` above the first element is `numbers[0]`, the second element is `numbers[1]`, and so on. For example, `numbers[0]=49`, assigns 49 to the first element of the array and `numbers[2]=52`, assigns 52 to the third element of the array.

Care must be taken when referencing the tenth element – it is `numbers[9]`, not `numbers[10]`.

The next program demonstrates a simple application of an array.

可以通过数组元素相对于数组中第一个元素的位置来引用数组中的每个元素。元素在数组中的位置称为数组的索引或下标。例如，数组numbers 中第一个元素的下标为0，最后一个元素的下标为9。注意，数组numbers的下标是从0至9，而不是从1至10。

可以使用数组名和方括号中的下标值来引用数组中的某个特定元素。在numbers 数组中，第一个元素是numbers[0]，第二个元素是numbers[1]，以此类推。例如，numbers[0]=49就是将49赋值给数组的第一个元素，而numbers[2]=52就是将52赋值给数组的第三个元素。

在引用数组的第十个元素时一定要小心，它是numbers[9]而不是numbers[10]。

Program Example P5A

```
1    // Program Example P5A
2    // Program to calculate the average age of ten people
3    // using an array.
4    #include <iostream>
5    using namespace std ;
6
7    int main( void )
8    {
9      int ages[10] ;
10     int total_age = 0 ;
11
12     cout << "Please enter the ages of ten people" << endl ;
13     // Input and total each age.
14     for ( int index = 0 ; index < 10 ; index ++ )
15     {
16       cin >> ages[index] ;
17       total_age += ages[index] ;
18     }
19     cout << "The average age is " << total_age / 10 << endl ;
20     return 0 ;
21   }
```

A sample run of this program is:

```
Please enter the ages of ten people
41
67
21
7
59
57
41
74
47
68
The average age is 48
```

The statement:

```
int ages[10] ;
```

defines ages as an array of ten integers. An array is defined by stating the type of its elements, its name, and the number of elements in the array. In general, the format is:

定义数组就是声明数组元素的类型、数组的名字、数组元素的个数。

```
data_type variable_name[number_of_elements] ;
```

For example:

```
float array_1[50] ; // An array of 50 floats.
char array_2[20] ; // An array of 20 characters.
```

The `for` loop in lines 14 to 18 is used to read in a value for each of the elements of the array and add them to `total_age`.

The variable `index` is used to hold the value of the *index* or *subscript* of the array. On the first pass through the loop, `index` is 0, and line 16 reads in a value for `ages[0]`, i.e. the first array element. Line 17 then adds `ages[0]` to `total_age`.

On the second pass through the loop, `index` has a value of 1, line 16 reads in a value for `ages[1]`, and line 17 adds `ages[1]` to `total_age`. The loop continues until the tenth element, `ages[9]`, is read in and added to `total_age`.

It is a common requirement in programming to find the minimum and maximum values in an array. The next program inputs ten ages and finds the youngest and the oldest. The average age is also computed.

第14行至第18行的for 循环用于读取数组中的每个元素的值，并将其累加到变量total_age中。

变量index用于保存数组的索引或下标。在第一次循环中，下标为0，第16行读取ages[0]的值，即数组第一个元素的值，第17行将ages[0]的值累加到变量total_age中。在第二次循环中，下标变为1，第16行读取ages[1]的值，第17行将ages[1]的值累加到变量total_age 中。循环将一直进行到将第10个元素ages[9]读入并累加到变量total_age中为止。

在编程中常常需要查找数组中的最小值和最大值。下面的程序输入10个人的年龄，然后查找其中最年轻和最年长的人，并计算平均年龄。

Program Example P5B

```
1   // Program Example P5B
2   // Program to read a series of ages and to find
3   // the youngest, the oldest, and the average.
4   #include <iostream>
5   using namespace std ;
6
7   int main( void )
8   {
9     const int SIZE = 10 ;
10    int ages[SIZE] ;
11    int i ;
12    int total_age = 0 ;
13    int youngest, oldest ;
14
15    cout << "Please enter " << SIZE << " ages" << endl;
16    // Input a value for each age.
17    for ( i = 0 ; i < SIZE ; i++ )
18    {
19      cin >> ages[i] ;
20      total_age += ages[i] ;
21    }
22
23    youngest = ages[0] ;
24    oldest = ages[0] ;
25
26    for ( i = 0 ; i < SIZE ; i ++ )
27    {
28      if ( ages[i] > oldest )
29      {
30        oldest = ages[i] ;
31      }
32      if ( ages[i] < youngest )
33      {
```

```
34        youngest = ages[i] ;
35      }
36   }
37
38   cout << "The youngest is " << youngest << endl ;
39   cout << "The oldest is " << oldest << endl ;
40   cout << "The average is " << total_age / SIZE << endl;
41   return 0 ;
42 }
```

This program starts by defining a constant integer SIZE on line 9. The const keyword is used in the definition of SIZE to specify that its value cannot be changed in the program. Although any valid identifier can be used for a constant, the identifier is usually written in uppercase. SIZE is known as a *symbolic constant*.

In line 9 SIZE is assigned a value of 10. The symbolic constant SIZE can now be used throughout the program in place of the number 10. Using a symbolic constant makes the program easier to modify. For example, to modify the above program to allow for twenty ages rather than ten, change line 9 to:

```
const int SIZE = 20 ;
```

The for loop in lines 17 to 21 reads in values into the array ages and totals them in total_age.

Lines 23 and 24 assign the first element of the array to the variables youngest and oldest. The for loop in lines 26 to 36 compares each element in the array with the values in the variables youngest and oldest. When an element larger than oldest is found, the value of this element is assigned to oldest. When an element is found that is less then youngest, this element is assigned to youngest. When the loop is completed, the smallest element of the array is in youngest and the largest is in oldest.

The variables oldest and youngest were initially assigned the value of the first element of the array. Any element of the array can be used, not necessarily the first.

5.1.2　Initialising an array

The next program demonstrates array initialisation by asking the user to enter a month and displaying the number of days in that month (leap years excepted).

Program Example P5C

```
1  // Program Example P5C
2  // Program to display the number of days in a month.
3  #include <iostream>
4  using namespace std ;
5
```

这个程序在开头的第9行定义了一个整型常量SIZE。关键字const用于定义SIZE，指定它的值不能在程序中被修改。尽管常量可以用任何有效的标识符来命名，但是通常将其全部大写。SIZE也称为符号常量。

第9行SIZE 被赋值为10，于是符号常量SIZE 就可以在整个程序中用于代替整数值10。使用符号常量使得程序更易于修改。

第23行和第24行语句将数组ages的第一个元素赋值给变量youngest和oldest。第26行至第36行的for循环将数组中的每个元素与变量youngest和oldest的值进行比较。当发现某个元素值大于oldest 时，就将该元素值赋值给oldest。当发现某个元素值小于youngest 时，就将该元素值赋值给youngest。循环结束后，存储在youngest中的元素就是数组中值最小的元素，而存储在oldest中的元素就是数组中值最大的元素。变量oldest和youngest被初始化为数组的第一个元素，但事实上可以将它们初始化为数组的任何一个元素，不一定非得是第一个元素。

下面是一个数组初始化的演示程序。程序先让用户输入月份的值，然后输出这个月份的天数（不考虑闰年）。

```
6   int main( void )
7   {
8      const int NO_OF_MONTHS = 12 ;
9      int days[NO_OF_MONTHS] =
10            { 31, 28, 31, 30, 31, 30, 31, 31, 30, 31, 30, 31 } ;
11     int month ;
12
13     cout << "Please enter a month (1 = Jan., 2 = Feb., etc.) " ;
14     do
15     {
16       cin >> month ;
17     }
18     while ( month < 1 || month > 12 ) ;
19
20     cout << endl << "The number of days in month " << month
21          << " is " << days[month-1] << endl ;
22     return 0 ;
23  }
```

A sample run of this program is:

```
Please enter a month (1 = Jan., 2 = Feb., etc.) 9
The number of days in month 9 is 30
```

Lines 9 and 10 of this program define and initialise an array days.
The initial values in the array are separated by commas and placed
between braces.

When the list of initial values is less than the number of elements in
the array, the remaining elements are initialised to 0. For example,

当初值个数少于数组元素个数时，其余的数组元素将被初始化为0。

```
float values[5] = { 2.3, 5.8, 1.3 } ;
```

initialises the first three elements with the values specified within the
braces. The remaining two elements of the array are initialised to 0.
If an array is defined without specifying the number of elements and
is initialised to a series of values, the number of elements in the array
is taken to be the same as the number of initial values. This means that

如果定义一个数组时对数组元素进行了初始化，但是没有指定数组元素个数，那么编译器将统计出花括号内提供的初值个数来作为数组元素的个数。

```
int numbers[] = { 0, 1, 2, 3, 4, 5, 6, 7, 8 } ;
```

and

```
int numbers[9] = { 0, 1, 2, 3, 4, 5, 6, 7, 8 } ;
```

are equivalent definitions of the array numbers.

5.1.3 Two-dimensional arrays

So far, only one-dimensional arrays, i.e. arrays with just one row of
elements, have been used. A two-dimensional array has more than
one row of elements. For example, to record the number of students
using one of five computer laboratories over a week, the data could
be recorded in a table of the form (see Table 5.1):

到目前为止，只用到了一维数组，即只有一行元素的数组。一个二维数组可以具有多于一行的元素。

Table 5.1　The number of students using computer laboratories

	Computer laboratory number				
	1	2	3	4	5
Day 1	120	215	145	156	139
Day 2	124	231	143	151	136
Day 3	119	234	139	147	135
Day 4	121	229	140	151	141
Day 5	110	199	138	120	130
Day 6	62	30	37	56	34
Day 7	12	18	11	16	13

This table has a row for each day of the week and a column for each computer laboratory. This is an example of a two-dimensional array. To define two-dimensional arrays, enclose each dimension of the array in brackets. For example,

```
int usage[7][5] ;
```

defines an integer array of seven rows and five columns.

To access an element of a two-dimensional array, you specify the row and the column. Note that the row number starts at 0 and ends at 6, and the column number starts at 0 and ends at 4. For example:

为了访问二维数组中的元素，必须指定元素的行号和列号。

```
usage[0][0] is 120  i.e. row 0, column 0
usage[0][4] is 139       row 0, column 4
usage[6][0] is  12       row 6, column 0
usage[6][4] is  13       row 6, column 4
```

The row number is in the first set of square brackets and the column number is in the second set of square brackets.

第一个方括号中的值代表行号，第二个方括号中的值代表列号。

The next program reads in the number of students using the five laboratories over seven days into a two-dimensional array `usage` and calculates the average usage for each laboratory. Run this program and study the code to see how it works.

Program Example P5D

```
1   // Program Example P5D
2   // Program to read in number of students using five computer labs
3   // over seven days and to display the average usage for each lab.
4   #include <iostream>
5   using namespace std ;
6
7   int main( void )
8   {
9     const int NO_OF_DAYS = 7 ;
10    const int NO_OF_LABS = 5 ;
11    int usage[NO_OF_DAYS][NO_OF_LABS] ;
12    int day, lab, total_usage, average ;
```

```
13
14   // Read each lab's usage for each day.
15   for ( day = 0 ; day < NO_OF_DAYS ; day++ )
16   {
17     cout << "Enter the usage for day " << ( day + 1 ) << endl ;
18     for ( lab = 0 ; lab < NO_OF_LABS ; lab++ )
19     {
20       cout << " Lab number " << ( lab + 1 ) << ' ' ;
21       cin >> usage[day][lab] ;
22     }
23   }
24
25   // Calculate the average usage for each laboratory.
26   for ( lab = 0 ; lab < NO_OF_LABS ; lab++ )
27   {
28     total_usage = 0 ;
29     for ( day = 0 ; day < NO_OF_DAYS ; day++ )
30     {
31       total_usage += usage[day][lab] ;
32     }
33     average = total_usage / NO_OF_DAYS ;
34     cout << endl << "Lab number " << ( lab+1 )
35            << " has an average usage of " << average << endl ;
36   }
37   return 0 ;
38 }
```

5.1.4   Initialising a two-dimensional array

A two-dimensional array, like a one-dimensional array, is initialised by enclosing the initial values in braces. For example,

和一维数组一样，二维数组也可以通过放置在花括号内的初值来对其进行初始化。

```
int vals[4][3] = { 4, 9, 5, 2, 11, 3, 21, 9, 32, 10, 1, 5 } ;
```

initialises the first row of vals with 4, 9, and 5. The second row is initialised with 2, 11, and 3. The third row is initialised with 21, 9, and 32 and the fourth row is initialised with 10, 1, and 5.
Readability is improved if you place the initial values of each row on a separate line, as follows:

将每行的初值单独写在不同的行上，可以提高程序的可读性。

```
int vals[4][3] = { 4,   9, 5,
                   2, 11, 3,
                  21,  9, 32,
                  10,  1, 5  } ;
```

Additional braces may also be used to separate the rows, as follows:

```
int vals[4][3] = { { 4,  9, 5  },
                   { 2, 11, 3  },
                   { 21, 9, 32 },
                   { 10, 1, 5 }  } ;
```

As with one-dimensional arrays, you can omit the number of rows and let the compiler calculate the number of rows from the initial values enclosed in the braces. Therefore you can rewrite the above definition of vals as:

和一维数组一样，可以省略数组的行数，这时编译器将根据花括号内提供的初值个数来计算数组的行数。

```
int vals[][3] = {  { 4,  9,  5  },
                   { 2,  11, 3  },
                   { 21, 9,  32 },
                   { 10, 1,  5  }  } ;
```

As with one-dimensional arrays, missing values are initialised to 0. For example, the definition:

和一维数组一样，未提供初值的数组元素将被初始化为0。

```
int vals[4][3] = {  { 4,  9 },
                    { 2 }  } ;
```

will result in

vals[0][0] = 4, vals[0][1] = 9 and vals[1][0] = 2

with all the remaining elements being 0. Note that the number of rows is required here or the compiler will assume it to be 2, as there are only two rows of initial values.

注意，这里数组的行数是不能省略的，否则编译器会将数组的第一维假定为2，因为这里只提供了两行初值。

5.1.5　Multi-dimensional arrays

You can define arrays with any number of dimensions. For example, if in program P5D you wanted to store the usage of the five laboratories for each day of the fifty-two weeks of a year, the array usage would be defined as:

还可以将数组定义为多维数组。例如，如果程序P5D要存储一年52个星期内每天5个实验室的使用情况，那么数组usage可以定义为一个三维数组。

```
const int NO_OF_WEEKS = 52 ;
const int NO_OF_DAYS = 7 ;
const NO_OF_LABS = 5 ;
int usage[NO_OF_WEEKS][NO_OF_DAYS][NO_OF_LABS] ;
```

The elements of this array are accessed by using three subscripts. For example,

访问该数组元素需要使用三个下标。

usage[0][2][4]

is the usage in the first week of day three in laboratory number five.

5.2　Structures（结构体）

5.2.1　Introduction

The items of information that make up an array have all the same data type (int, float etc.) and are logically related in some way. For example, a student's test scores may be integer values that are logically related to the student. In this case it makes sense to store the test scores together in an array. In short, arrays are suitable for storing sets of homogeneous data.

构成一个数组的各信息项在某种程度上都是逻辑相关的，并且具有相同的数据类型（整型、实型等）。例如，一个学生的考试成绩与其他学生是逻辑相关的，并且都是整型值。在这种情况下，将考试成绩存储在一个数组中是合理的。简而言之，数组适合于存储同种类型的数据集。

However, there are items of information that are logically related but each item may have a different data type. A student's number and test scores, for example, are logically related to the student, but the number may be an integer, while the test scores may be floating-point values.

Logically related items of information that may have different data types can be combined into a *structure*. Unlike an array, the data items in a structure may be of different types.

然而，还有一些逻辑相关但数据类型不同的信息项，例如，学生的学号和考试成绩是逻辑相关的，但是学号可能是整型数据，而考试成绩可能是浮点型数据。

结构体可以表示逻辑相关但数据类型不同的数据项。与数组不同的是，结构体中的数据项的数据类型可以是不同的。

5.2.2   Declaring a structure

The first step in defining a structure is to declare a *structure template*:

定义结构体的第一步是声明一个结构体模板。

```
struct student_rec
{
  int number ; // Student number.
  float scores[5] ; // Scores on five tests.
} ;
```

A structure template consists of the reserved keyword `struct` followed by the name of the structure. The name of the structure is known as the *structure tag*. In the example above, `student_rec` is a structure tag.

After the structure tag, each item within the structure is declared within the braces { and }. Each item in a structure is called a *structure member*. A structure member has a name and a data type. Any name can be used for a structure member, provided it is a valid C++ identifier.

Declaring a structure template does not allocate memory to the structure. All that has been done at this stage is to define a new data type consisting of other previously defined data types. Once you have defined the new data type you can then define variables with that type. For example,

一个结构体模板是由关键字struct及其后的结构体名字组成的。结构体的名字也称为结构体标记。在结构体标记的后面，结构体中的各信息项是在一对花括号{和}内声明的。结构体中的信息项称为结构体成员。每个结构体成员都有一个名字和相应的数据类型。结构体成员可以任意命名，只要是有效的C++标识符即可。声明结构体模板无须为其分配内存。声明结构体模板的目的是定义一个由已有数据类型构成的新的数据类型。一旦定义完这个新的数据类型，就可以用该类型来定义变量。

```
struct student_rec student1, student2 ;
```

defines the variables `student1` and `student2` to be of the type `struct student_rec`. Both `student1` and `student2` are structure variables with two structure members, i.e. `number` and `scores`.

student1:

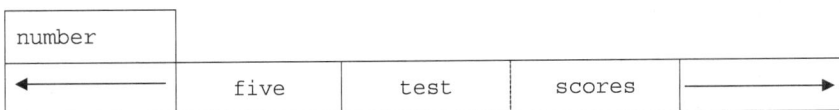

student2:

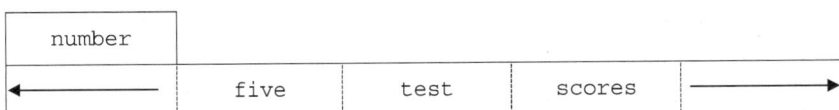

The members of a structure variable can be accessed with the member selection operator "." (a dot). For example, we can assign values to the member `number` of the variables `student1` and `student2` with the statements:

```
student1.number = 1234 ; student2.number = 13731 ;
```

The two variables `student1`.number and `student2`.number are used in the same way as any other integer variable.

The next program inputs values for each member of a structure and displays it on the screen.

结构体变量的成员可以通过成员选择运算符 "."（一个圆点）来访问。

Program Example P5E

```
1    // Program Example P5E
2    // Introduction to structures: assigning values to structure
3    // members.
4    #include <iostream>
5    #include <iomanip>
6    using namespace std ;
7
8    int main( void )
9    {
10     int i ;
11
12     // Declare the structure template.
13     struct student_rec
14     {
15       // Declare the members of the structure.
16       int number ;
17       float scores[5] ;
18     } ;
19
20     // Define two variables having the type struct student_rec.
21     struct student_rec student1, student2 ;
22
23     // Read in values for the members of student1.
24     cout << "Number: " ;
25     cin >> student1.number ;
26     cout << "Five test scores: " ;
27
28     for ( i= 0 ; i < 5 ; i++ )
29       cin >> student1.scores[i] ;
30
31     // Now assign values to the members of student2.
32     // The assignments are not meant to be meaningful and
33     // are for demonstration purposes only.
34     student2.number = student1.number + 1 ;
35     for ( i = 0 ; i < 5 ; i++ )
36       student2.scores[i] = 0 ;
37
38     // Display the values in the members of student1.
```

```
39    cout << endl << "The values in student1 are:" ;
40    cout << endl << "Number is " << student1.number ;
41    cout << endl << "Scores are:" ;
42    cout << fixed << setprecision( 1 ) ;
43    for ( i = 0 ; i < 5 ; i++ )
44      cout << setw(5) << student1.scores[i] ;
45
46    // Display the values in the members of student2.
47    cout << endl << endl << "The values in student2 are:" ;
48    cout << endl << "Number is " << student2.number ;
49    cout << endl << "Scores are:" ;
50    for ( i = 0 ; i < 5 ; i++ )
51      cout << setw(5) << student2.scores[i] ;
52
53    cout << endl ;
54    return 0 ;
55  }
```

A sample run of this program is shown below.

```
Number: 1234
Five test scores: 4.5   6.0   5.5   6.5   7.5

The values in student1 are:
Number is 1234
Scores are: 4.5   6.0   5.5   6.5   7.5

The values in student2 are:
Number is 1235
Scores are: 0.0   0.0   0.0   0.0   0.0
```

The structure tag `student_rec` in line 13 of program P5E is optional when the structure template and the structure variables are defined together:

当结构体模板和结构体变量一起定义时，程序P5E第13行的结构体标记student_rec 是可选的。

```
// Declaring a structure template without a structure tag.
struct     // No tag name after struct.
{
  int number ;
  float scores[5] ;
} student1, student2 ; // Variables follow immediately after the }.
```

5.2.3   Initialising a structure variable

The members of a structure variable can be initialised by placing their initial values in braces.
Example:

结构体变量的成员可以通过将其初值置于花括号之内来进行初始化。

```
struct student_rec
{
  int number ;
  int scores[5] ;
} ;
```

```
struct student_rec student = { 1234,
                               { 50, 60, 45, 65, 75 }
                             } ;
```

The first member of the structure (`student.number`) is initialised to `1234`, the second member (`student.scores`) is an integer array and is initialised to the values enclosed in the inner set of braces. The initial values are on separate lines for visual purposes only, making it easy to relate a structure member with its initial value.

5.2.4　Nested structures

A *nested structure* is a structure that contains another structure as one of its members. For example, a company personnel record might consist of, among other things, the employees' date of birth and date of joining the company. Both these dates can be represented by a structure with members day, month and year.

First declare the structure template for a date as follows:

```
struct date  // Structure template for a date.
{
  int day ;
  int month ;
  int year ;
} ;
```

Next, the template for the structure `personnel` is declared in terms of the previously declared structure template `date`.

```
struct personnel  // Structure template for an employee.
{
  int number ;  // Employee number.
              // and various other structure members, e.g. pay.
  struct date dob ;   // The data type of dob is struct date.
  struct date joined ; // joined is also of type struct date.
} ;
```

Finally, define a variable `person` of the type `struct personnel`, as in:

```
struct personnel person ;
```

Graphically, the personnel structure looks like this:

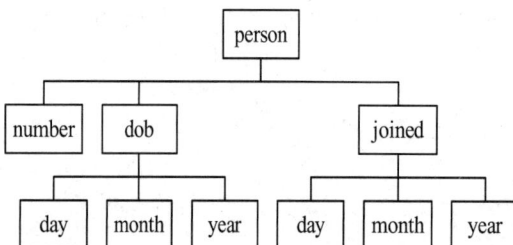

结构体的第一个成员（student.number）初始化为1234，第二个成员（student.scores）是一个整型数组，初始化为花括号内的数值。为了提高程序的可读性，这些初值分别独立成行，使其更容易与相对应的结构体成员关联在一起。

嵌套的结构体就是在一个结构体内包含另一个结构体作为其成员。例如，一个公司的个人记录中可能包括雇员的出生日期及进入公司工作的日期。这两个日期都用一个拥有年、月、日成员的结构体来表示。

The expressions:

```
person.dob and person.joined
```

will access the date of birth and date of joining members, respectively. Furthermore,

```
person.dob.day person.dob.month and person.dob.year
```

will access the day, month and year of birth, respectively. Similarly,

```
person.joined.day   person.joined.month
and person.joined.year
```

will access the day, month and year of the date the person joined the company.

5.3   The `typedef` statement（typedef 语句）

`typedef` allows you to define a synonym for a built-in or a programmer-defined data type.

From the personnel example on the previous pages, we had the following structure templates:

typedef 允许程序员为系统内置的或者程序员自定义的数据类型名定义一个同义词。

```
struct date  // Structure template for a date.
{
  int day ;
  int month ;
  int year ;
} ;


struct personnel  // Structure template for an employee.
{
  int number ;  // Employee number.
               // and various other structure members, e.g. pay.
  struct date dob ;    // The data type of dob is struct date.
  struct date joined ; // joined is also of type struct date.
} ;
```

The following statement uses `typedef` to define a synonym DATE for `struct date`:

```
typedef struct date DATE ;
```

The personnel structure template can now be written as:

```
struct personnel
{
  int number ; // Employee number
               // and various other structure members.
  DATE dob ;
  DATE joined ;
};
```

Going a step further, a synonym EMPLOYEE can be written for struct personnel:

```
typedef struct personnel EMPLOYEE ;
```

The variable person can now be conveniently defined as:

```
EMPLOYEE person ;
```

5.4　Arrays of structures（结构体数组）

Continuing with the personnel example used in the previous sections:

```
struct personnel persons[5] ;
```

or

```
EMPLOYEE persons[5] ;
```

defines a five-element array persons. Each element of this array is of the type struct personnel, with members number, dob and joined. The members dob and joined are themselves structures and have members day, month and year.

Note that persons[0].number will access the employee number of the first employee and persons[4].joined.year will access the year of joining of the fifth employee.

下面的语句定义了一个有5个元素的数组persons。数组的每个元素都是拥有3个成员number、dob和joined的struct personnel的结构体。成员dob和joined本身又是拥有日、月和年3个成员的结构体。注意，persons[0].number访问的是第一个雇员的雇员号码，persons[4]. joined.year 访问的是第五个雇员进入公司工作的年份。

5.5　Enumerated data types（枚举数据类型）

An enumerated data type is used to describe a set of integer values. For example:

枚举数据类型用于描述一组整型的数值。

```
enum response { no, yes, none } ;
enum response answer ;
```

These statements declare the data type response to have one of three possible values: no, yes, or none. The variable answer is defined as an enumerated variable of type response. This is similar to the way in which a structure template and a structure variable are defined.

The names enclosed in the braces { and } are integer constants. The first name (no) has a value of 0, the second name (yes) has a value of 1, and the third name (none) has a value of 2. The variable answer can be assigned any of the possible values: no, yes, or none.

For example:

上述语句声明了一个名为response的枚举数据类型，它的可能取值为：no、yes或none。变量answer定义为response枚举类型的变量。这种定义形式和结构体模板及结构体变量的定义形式很相似。

花括号内的名字都是整型常量。第一个名字（no）的值为0，第二个名字（yes）的值为1，第三个名字（none）的值为2。可以用no、yes或none中的任意一个值给变量answer赋值。

```
answer = none ;
```

or

```
answer = no ;
```

The variable answer can also be used in an if statement. For example:

```
if ( answer == yes )
{
  // statement(s)
}
```

The purpose of the enumerated data type is to improve the readability of the program. In the example above, using `yes`, `no` and `none` rather than 0, 1 and 2 makes the program more readable.

In this example, `response` is called the *enumeration tag*. Like a structure tag, the enumeration tag is optional when the enumerated data type and the enumerated variables are defined together.

For example, the variable `answer` could also be defined as:

```
enum { no, yes, none } answer ;
```

Arrays can also be used. For example:

```
enum response answers[200] ;
```

Values other than 0, 1, and 2 can also be used. For example:

```
enum response { no = -1, yes = 1, none = 0 } ;
```

To add another possible value to `response`, include the new value within the braces. For example:

```
enum response { no = -1, yes = 1, none = 0, unsure = 2 } ;
```

使用枚举数据类型的目的是提高程序的可读性。

在这个例子中，response称为枚举标记。和结构体标记类似，当枚举数据类型和枚举变量放在一起定义时，枚举标记是可选的。

Programming pitfalls

1. The number of elements in each dimension of an array are placed between brackets [ ] and not between parentheses ( ).
2. The range of a subscript is 0 to the number of elements in an array less one. It is a common error to define an array with, for example, ten elements and then attempt to use a subscript value of 10. The subscripts in this case range from 0 to 9.

    For example:

```
int i, a[10] ;
for ( i = 0 ; i <= 10 ; i++ )
  a[i] = 0 ;
```

This may cause an infinite loop. When `i` is `10`, `a[i]` is assigned `0`. However, `a[10]` does not exist, so `0` is stored in the memory location immediately after `a[9]`. If the variable `i` happens to be stored after `a[9]`, then `i` becomes `0`, and so the loop starts again.

3. You cannot compare structure variables in an `if` statement, even if they have the same structure template. For example, if `s1` and `s2` are defined as:

```
struct
{
  int a ;
  int b ;
  float c ;
} s1, s2 ;
```

The two variables `s1` and `s2` cannot be tested for equality with the statement:

```
if ( s1 == s2 )    // Invalid.
```

To test `s1` and `s2` for equality you must test each member of each structure for equality, as in the statement

```
if ( s1.a == s2.a && s1.b == s2.b && s1.c == s2.c )
```

1. 应将数组的每一维的元素个数放在方括号之内，而非圆括号之内。
2. 数组下标的取值范围是从0开始到数组的元素个数减1。

3. 即使结构体的模板是相同的，也不能在if语句中对两个结构体变量进行比较。

Quick syntax reference

	Syntax	Examples
Defining arrays	`type array[d1][d2]...[dn] ;` Dimensions `d1,d2...dn` are integer constants.	`int a[10] ;` `float b[5][9] ;`

Array subscripts	`array[i1][i2]...[in]`
	indexes or subscripts `i1,i2...in` are
	integer constants or variables.
		`a[0] // 1st  element.`
		`a[9] // 10th element.`
		`b[0][0] // Row 1, col 1.`
		`b[4][8] // Row 5, col 9.`

(cont.)

	Syntax	Examples
Declaring a structure template	```struct structure_tag { type variable1 ; type variable2 ; ... } ;```	```struct date { int day ; int month ; int year ; } ;```
Defining structure variables	```struct structure_tag variable₁, variable₂, ... ;```	```struct date dob ;```
Accessing structure members	Member selection operator. (Dot operator)	```dob.day ;```

Exercises

1. What are the subscript ranges of the following arrays?

   (a) `int array1[6] ;`

   (b) `float array2[] = { 1.3, 2.9, 11.8, 0 } ;`

   (c) `int array3[6][3] ;`

   (d) `int array4[][4] = { { 6, 2, 1, 3 }, { 7, 3, 8, 1 } } ;`

2. Write statements to define each of the following:

   (a) a one-dimensional array of floating-point numbers with ten elements

   (b) a one-dimensional array of characters with five elements

   (c) a two-dimensional array of integers with seven rows and eight columns

   (d) a 10 by 5 two-dimensional array of double precision numbers

   (e) a 10 by 8 by 15 three-dimensional array of integers.

3. What is the output from the following program?

```
int i, c1 = 0, c2 = 0 ;
int a[] = { 6, 7, 3, 13, 11, 5, 1, 15, 9, 4 } ;
for ( i = 0 ; i < 10 ; i++ )
{
  if ( i % 2 == 0 )
    c1++ ;
  if ( a[i] % 2 == 0 )
    c2++ ;
}
cout << "c1 = " << c1 << " c2 = " << c2 ;
```

4. Write a program to read in fifteen numbers from the keyboard and display them as follows:

   (a) each number on a separate line

   (b) on one line, each number separated by a single space

   (c) as in (b) but in the reverse order to which they were input.

5. Write a program to input numbers to two one-dimensional arrays, each having five elements, and display the result of multiplying corresponding elements together.

6. The number of customers entering a shop per hour is recorded for each of the nine hours the shop is open. Write a program to display a report of the form:

```
   Time            Number of customers    Percentage of total
 9:00 - 10:00             153                    10
10:00 - 11:00             189                    12
 ...                      ...                    ...
17:00 - 18:00             135                     9
```

7. The following two arrays represent the fixed and variable costs involved in producing each of eight items:

```
float fixed[] = { 11.31, 12.12, 13.67, 11.91, 12.30, 11.8, 11.00, 12.00 } ;

float variable[] = { 1.12, 1.13, 3.14, 1.35, 2.20, 1.28, 1.00, 2.10 } ;
```

Write a program to input an item number in the range 1 to 8 along with the number of units produced. The program should then display the total cost of producing that number of units, where the total cost is the sum of the fixed and variable costs.

（fixed costs：固定成本；variable costs：可变成本；total costs：总成本）

8. Use two `for` loops to set all the diagonal elements of a 9 by 9 integer array to 1 and all the elements not on a diagonal to 0.

（diagonal elements：对角线上的元素）

9. Write a program to input values to a 4 by 5 array, search the array for values that are less than 0 and display these values along with their row and column indices.

10. Write a program to input ten integer values into an array `unsorted`. Your program should then loop through `unsorted` ten times, selecting the lowest value during each pass. For each pass through the loop, the element in `unsorted` containing the lowest value is replaced with a large value (e.g. 9999) after copying it into the next available element of another integer array `sorted`.

This is illustrated below:

`unsorted` at the start: 14 22 67 31 89 11 42 35 65 49

    `sorted` at the start:

`unsorted` after the first pass: 14 22 67 31 89 9999 42 35 65 49

    `sorted` after the first pass: 11

`unsorted` after the second pass: 9999 22 67 31 89 9999 42 35 65 49

    `sorted` after the second pass: 11   14

etc.

Display the values in `sorted`. (Hint: see program P5B to determine the smallest value.)

11. In a magic square the rows, columns and diagonals all have the same sum. For example:

17	24	1	8	15
23	5	7	14	16
4	6	13	20	22
10	12	19	21	3
11	18	25	2	9

and

4	9	2
3	5	7
8	1	6

Write a program to read in a two-dimensional integer array and check if it is a magic square.
（magic square：幻方）

12. It is required to scale a ten element floating-point array a so that the maximum element in the array becomes 1, the minimum element becomes 0 and the other elements are scaled to between 0 and 1 according to their values. The following statement computes the scaled value of element a[i]

```
a[i] = ( a[i] -min_value ) / ( max_value - min_value ) ;
```

where min_value and max_value are the minimum and maximum values in the array a.

Write a program to read in values for the elements of a and display the elements of a scaled to values between 0 and 1.（scale：比例变换）

13. Write a structure template for each of the following:

    (a) the time of day using the twenty-hour format, i.e. hours, minutes and seconds

    (b) a playing card, such as the five of diamonds or the three of spades. The structure members will be an integer to represent the card value and a character to represent the suit

    (c) a transaction record consisting of a transaction type (one character), the date of the transaction (three integers), and the amount of the transaction (floating point)

    (d) the longitude and latitude co-ordinates of a geographical position consisting of degrees (integer), minutes (integer) and direction ('N', 'S', 'E' or 'W')

    （structure template：结构体模板；playing card：扑克牌；diamonds：方块；spades：黑桃；card value：牌面；suit：花色；transaction record：交易记录；longitude：经度；latitude：纬度）

14. Given the following:

```
struct stock_record
{
    int stock_number ;
    float price ;
    int quantity_in_stock ;
} ;

struct stock_record stock_item ;
```

    write statements to

    (a) assign a value to each member of stock_item

    (b) input a value to each member of stock_item

    (c) display the value of each member of stock_item

15. Create an enumerated data type for each of the following:

    (a) the days of the week: Monday, Tuesday, Wednesday, and so on

    (b) the months of the year

    (c) monetary denominations

    (d) the suits in a pack of cards

    (e) the points on a compass.

    （compass：指南针）

Chapter Six
Strings
第 6 章　字　符　串

6.1　C-strings（C 风格字符串）

In the C programming language, a string is an array of characters (elements of type `char`) with the null character `'\0'` in the last element of the array. Because it comes from C, this type of string is called a C-string. C-strings are used in many instances in C++.

A C-string is an array of characters and can be initialised in the same way as any other array. For example:

```
char greetings[6] = {'H', 'e', 'l', 'l', 'o', '\0'} ;
```

This statement initialises a six-element `char` array `greetings` with the character constants that spell the word "Hello". Note that the last element of the array `greetings` is the null character (`'\0'`). Without the null character `greetings` is a character array but not a proper C-string.

An easier way to initialise `greetings` is:

```
char greetings[] = "Hello" ;
```

This statement shows that it is not necessary to specify the number of characters in the array or to initialise each element individually. In the definition of the array `greetings`, the compiler determines the size of the array by the number of characters in the array plus 1 (for the null character `'\0'`).

A sequence of characters enclosed in double quotation marks is called a *string literal*. The compiler automatically inserts the null character `'\0'` after the last character of a string literal. The string literal "Hello", for example, actually contains six, rather than five characters.

Just as with other array types, the individual elements of `greetings` can be accessed using subscripts:

```
greetings[0] is 'H'
greetings[1] is 'e'
```

If you specify the size of the array and the string is shorter than this size, the remaining elements of the array are initialised with the null

在C语言中，字符数组是元素为字符型的数组，字符串是以空字符'\0'作为数组最后一个元素的字符数组。由于C++语言起源于C语言，因此这种类型的字符串称为C风格字符串（或C字串）。C++中的很多实例都用到了C风格字符串。

C风格字符串是字符数组，因此可以采用和其他数组一样的方式进行初始化。

我们不必指定数组中字符的个数，也不必单独为数组中的每个元素进行初始化。在定义数组greetings 时，编译器会根据字符的个数来确定数组的大小，由于字符数组的最后一个元素为'\0'，因此数组的大小为字符的个数加1。用双引号引起来的字符序列，称为一个字符串字。在字符串字的末尾，编译器会自动添加'\0'。

如果指定了数组的大小，而字符串的长度又小于数组的大小，那

character '\0'. For example:

```
char greetings[9] = "Hello" ;
```

initialises greetings to:

'H'	'e'	'l'	'l'	'o'	'\0'	'\0'	'\0'	'\0'

To include a double quote inside a string precede the quote with a back slash (\ ). For example:

```
char greetings[] = "\"Hello\", I said." ;
cout << greetings ;
```

will display

```
"Hello", I said.
```

The \" is an example of an escape sequence. Further examples of escape sequences are given in appendix F.

The newline ('\n') escape sequence can be used in place of endl to advance to a new line. For example,

```
cout << "some text\n" ;
```

is equivalent to

```
cout << "some text" << endl ;
```

If a string is too long to fit onto a single program line, it can be broken up into smaller segments. For example:

```
char long_string[] = "This is the first half of the string "
                     " and this is the second half." ;
```

么这个数组的其余元素都将被初始化为'\0'。

为了在一个字符串中包含一个双引号，必须在这个双引号的前面加上一个反斜杠（\）。

6.2　C-string input and output
（C 风格字符串的输入和输出）

C-strings may be read and displayed in much the same way as for any other data. There is, however, an important consideration to keep in mind when using C-strings and that is to allow for sufficient storage to hold the string. Remember that the number of elements in the char array must be one more than the number of characters in the string.

The next program is a simple demonstration of C-string input and output. The program inputs a name from the keyboard and displays it on the screen.

如果一个字符串太长，不能写在一行中，那么可以把它拆分成几个小的片段，写在不同的行中。

像其他类型的数据一样，C风格字符串可以用很多方式进行读取和显示。但是，在使用C风格字符串时，有一点要牢记，就是要为字符串留有足够的存储空间。请记住，字符数组的大小必须比字符串中的字符数多1。

Program Example P6A

```
1  // Program Example P6A
2  // Program to read in a string of characters from the keyboard
3  // and to display it on the screen.
4  #include <iostream>
```

```
5   using namespace std ;
6
7   int main( void )
8   {
9     const int MAX_CHARACTERS = 10 ;
10    char first_name[ MAX_CHARACTERS + 1 ] ;
11
12    cout << "Enter your first name (maximum "
13        << MAX_CHARACTERS << " characters) " ;
14    cin >> first_name ;
15    cout << "Hello " << first_name << endl ;
16    return 0 ;
17  }
```

A maximum of 10 characters plus 1 for '\0' is stored in first_name.

The following is a sample run of this program:

```
Enter your first name (maximum 10 characters) John
Hello John
```

Another sample run of this program is:

```
Enter your first name (maximum 10 characters) John Paul
Hello John
```

What happened to Paul? Why was it not displayed?

The extraction operator >> read the characters up to, but not including, the space character after John. The remaining characters (Paul) are left in the input stream and are not extracted.

If the user does not follow the instructions and types in more than ten characters, the array first_name will overflow, the excess characters will overwrite other areas of memory, and the program will probably malfunction. To allow for this possibility and also to allow for whitespace characters in the input, the getline() function can be used.

为什么没有显示Paul 呢？
流提取运算符>>读取字符直到遇到John后的空格符时为止（但是空格符没有被读取）。于是，空格后面的字符（Paul）就被留在了输入流中，没有被读取。
如果用户没有按照圆括号中的指示，输入的字符多于10个，那么数组first_name就会溢出，多出的那些字符就会改写内存的其他区域，程序很可能就会出错。考虑到这种可能性及为了允许输入空格，可以使用函数getline()。

Program Example P6B

```
1   // Program Example P6B
2   // Program to read in a string of characters containing whitespaces
3   // from the keyboard and to display it on the screen.
4   #include <iostream>
5   using namespace std ;
6
7   int main( void )
8   {
9     const int MAX_CHARACTERS = 10 ;
10    char first_name[ MAX_CHARACTERS + 1 ] ;
11
12    cout << "Enter your first name(maximum "
13        << MAX_CHARACTERS << " characters) " ;
14    cin.getline( first_name, MAX_CHARACTERS + 1, '\n' ) ;
15    cout << "Hello " << first_name << endl ;
16    return 0 ;
17  }
```

Same number as the dimension of the array.

Line 14 reads characters from `cin` until either the user presses the Enter (or newline) key `'\n'` or 10 characters have been read. The newline character `'\n'` is called the *delimiter* and signifies the end of the input from `cin` to `first_name`. The delimiter can be any character.

If the delimiter is omitted, it is assumed to be `'\n'` and so line 14 can also be written as:

```
cin.getline( first_name, MAX_CHARACTERS + 1 ) ;
```

The characters are stored in the character array `first_name` with the null character `'\0'` automatically added by `getline`.

Since `'\0'` is automatically added to the end of an extracted string, the maximum number of characters extracted from the stream by `getline` is 1 less than specified. The number of characters specified should not be more than the number of elements in the character array or the array will overflow into other areas of memory.

The function `getline()` is a member function of the input stream object `cin`. It is more powerful than the extraction operator `>>`, since it allows the input of characters to stop after a specified delimiter is read. There is another problem that can arise when inputting data from the keyboard. The following program inputs a student number and a name from the keyboard and displays it on the screen.

换行符'\n'称为定界符，表示从输入流cin读取字符，直到数组first_name的末尾。其他字符也可以作为定界符。

若定界符被省略，则将其假定为'\n'。

当字符被存储在字符数组first_name中之后，函数getline会自动在其末尾添加'\0'。

由于'\0'是自动添加到字符串末尾的，因此用getline从输入流中读取的最大字符个数会比指定的长度少1。所以，指定的字符个数不能大于字符数组的大小，否则数组会溢出。

函数getline( )是输入流对象cin的一个成员函数。它的功能要比流提取运算符>>的强大，因为它在读取定界符以后，将终止字符的输入。

Program Example P6C

```
1   // Program Example P6C
2   // Program to read in a student number and name from the keyboard.
3   #include <iostream>
4   using namespace std ;
5
6   int main( void )
7   {
8     const int MAX_CHARACTERS = 20 ;
9     char student_name[ MAX_CHARACTERS + 1 ] ;
10    int student_number ;
11
12    cout << "Enter student number: " ;
13    cin >> student_number ;
14    cout << "Enter student first name and surname (maximum "
15        << MAX_CHARACTERS << " characters) " ;
16    cin.getline( student_name, MAX_CHARACTERS + 1 ) ;
17    cout << endl << "Data Entered:" << endl
18        << "Student Number: " << student_number << endl
19        << "Student Name: " << student_name << endl ;
20    return 0 ;
21  }
```

A sample run of this program follows:

```
Enter student number: 12345 ◄────── Enter pressed here is read as the name.
Enter student first name and surname (maximum 20 characters)
Data Entered:
Student Number: 12345
Student Name: ◄
```

The prompt on line 14 is displayed correctly but line 16 seems to be skipped and no name is read in from the keyboard.

The problem is as follows:

- Line 13 stops reading into the numeric variable student_number as soon as a non-numeric character is read.
- The Enter key that is pressed after typing 12345 is left in the input stream, which is then read by getline() on line 16 into student_name.

That's why the program didn't wait for the user to enter a name and why line 19 displayed an empty string.

One solution to this problem is to read the newline character '\n' into a 'dummy' character variable. For example, after line 13 include the line:

```
char dummy ; cin.get( dummy ) ;
```

This will work provided the user doesn't type any superfluous characters (e.g. spaces) after the student number before pressing Enter. If a space is typed after the student number, then it will be read into dummy and the '\n' will remain in the input stream.

The complete solution to the problem is to discard or ignore all characters in the input stream up to and including the newline character '\n'. Only then should getline() be used to read data into student_name.

This can be accomplished by using the ignore() function, as shown in the next program.

第14行中的提示信息被正确地显示了，但是第16行好像被跳过去了，并没有从键盘读取名字。问题出在：

- 第13行从输入流将数据读入数值变量student_number中，直到读入一个非数值型数据时为止。
- 在输入12345 之后输入的回车符留在了输入流中，它被第16行的函数getline( )读入student_name 中。

这就是为什么程序没有等待用户输入一个名字，以及为什么第19行显示了一个空字符串的原因。

对于这个问题的一种解决办法是把'\n'读入一个"哑的/虚拟的"字符型变量dummy中。

如果用户在输入学号之后、回车符之前不再输入任何多余的字符（如空格），那么这个方法是可行的。但如果在学号之后多输入了一个空格，那么这个空格就会被读入变量dummy中，而'\n'则被继续留在输入流中。

一个完整的解决方案是丢弃或者忽略输入流中包括'\n'在内的所有多余的字符。只有这样，getline( )才能将数据读入student_name中。使用函数ignore( )可以实现这个功能，如下面程序所示。

Program Example P6D

```
1  // Program Example P6D
2  // Corrected program to read in a student number
3  // and name from the keyboard.
4  #include <iostream>
5  using namespace std ;
6
7  int main( void )
8  {
9    const int MAX_CHARACTERS = 20 ;
10   char student_name[ MAX_CHARACTERS + 1 ] ;
11   int student_number ;
12
```

```
13    cout << "Enter student number: " ;
14    cin >> student_number ;
15    cout << "Enter student first name and surname (maximum "
16         << MAX_CHARACTERS << " characters) ";
17    cin.ignore( 80,'\n' ) ;
18    cin.getline( student_name, MAX_CHARACTERS + 1 ) ;
19    cout << endl << "Data Entered:" << endl
20         << "Student Number: "<< student_number << endl
21         << "Student Name: "<< student_name << endl ;
22    return 0 ;
23  }
```

Execution of cin.ignore(80, '\n') on line 17 removes at most 80 characters up to and including the newline character '\n' from the input stream. The maximum number of characters to remove is usually set to 80 for keyboard input, but this may be changed if required.

执行第17行中的cin.ignore (80, '\n'), 可以从输入流中删除包括'\n'在内的最多80个字符。对于键盘输入来说，可删除的最大字符数通常设置为80，但如果需要也可以改变这个值。

6.3   Accessing individual characters of a C-string
（访问C风格字符串中的单个字符）

As a C-string is an array of characters, each character of a C-string can be accessed using an index. The next program displays each character of the string "Hello" on separate lines.

由于C风格字符串是一个字符数组，因此可以通过下标来访问C风格字符串的每个字符。

Program Example P6E

```
1   // Program Example P6E
2   // Program to display each character of a C-string on a new line.
3   #include <iostream>
4   using namespace std ;
5
6   int main( void )
7   {
8     char greetings[6] = "Hello" ;
9     // Display each character of greetings on a new line.
10    for ( int i = 0 ; i < 5 ; i++ )
11    cout << greetings[i] << endl ;
12    return 0 ;
13  }
```

This program will display the following lines:

```
H
e
l
l
o
```

6.4   C-string functions（C 风格字符串函数）

C++ has inherited a library of C-string functions from the C programming language. To use any of these functions it is necessary to include the

C++继承了C语言中的C风格字符串函数库。

following line in the program:

```
#include <cstring>
```

6.4.1 Finding the length of a C-string

The standard library function `strlen()` returns the number of characters in a C-string, excluding the null character `'\0'`.
For example:

标准库函数strlen( )返回一个C风格字符串中字符的个数，不包含'\0'。

```
char name1[] = "Sharon" ;
char name2[10] = "Mark" ;
int len ;
len = strlen( name1 ) ;
cout << setw( 3 ) << strlen( name1 )
     << setw( 3 ) << strlen( name2 )
     << setw( 3 ) << strlen( "Rob" )
     << setw( 3 ) << len ;
```

This will display:

```
6   4   3   6
```

The general format of the `strlen()` function is:

```
len = strlen( str )
```

where `str` is a null-terminated string and `len` is an integer. (A null-terminated string is a string of characters ending with the null character `'\0'`.)

6.4.2 Copying a C-string

The C-string copy function, `strcpy( str1, str2 )`, copies the contents of a C-string `str2` to another C-string `str1`.
For example:

C风格字符串复制函数strcpy (str1, str2)的功能是把字符串str2中的内容复制到字符串str1中。

```
char name1[] = "Sharon" ;
char name2[10] = "Mark" ;
// Copy the contents of name1 to name2.
strcpy( name2, name1 ) ;
// Restore the original name.
strcpy( name2, "Mark" ) ;
```

The general format of `strcpy()` is:

```
strcpy( destination, source ) ;
```

where the source string is copied to the destination string. The source string must be null-terminated, i.e. a `'\0'` must be at the end of the string. The `strcpy()` function assumes that the destination string is big enough to hold the string being copied to it. No checking is performed by the compiler, so beware!

这里，source字符串中的内容被复制到destination字符串中。source字符串必须以'\0'结束。函数strcpy( )假定destination字符串有足够的空间来存储要复制的字符串内容。编译器不会检查其存储空间的大小，所以要小心。

6.4.3   C-string concatenation

The function strcat(str1, str2) concatenates a C-string str2 to the end of the C-string str1. Both str1 and str2 must be null-terminated. Enough memory must be allocated to str1 to hold the result of the concatenation.

For example:

函数strcat(str1, str2)的功能是把字符串str2拼接到字符串str1的后面。其中str1和str2都必须以'\0'结束，同时str1要有足够的空间来存储拼接后的字符串。

```
char str1[15] = "first & " ;
char str2[] = "second" ;
strcat( str1, str2 ) ; // str1 is now "first & second".
                       // str2 is unchanged.
```

6.4.4   Comparing C-strings

The function strcmp(str1, str2) compares two null-terminated C-strings str1 and str2. This function returns a negative value if the string in str1 is less than the string in str2, 0 if the string in str1 is equal to the string in str2, and a positive value if the string in str1 is greater than the string in str2.

The next program demonstrates strcmp().

函数strcmp(str1, str2)的功能是比较两个以'\0'结束的字符串str1和str2的大小。如果str1小于str2，则函数返回一个负值；如果str1等于str2，则函数返回0；如果str1大于str2，则函数返回一个正值。

Program Example P6F

```
1   // Program Example P6F
2   // Program to demonstrate strcmp() for comparing C-strings.
3   #include <iostream>
4   #include <cstring>
5   using namespace std ;
6
7   int main( void )
8   {
9     char password[7] = "secret" ;
10    char user_input[81] ;
11    cout << "Enter Password: " ;
12    cin >> user_input ;
13    if ( strcmp( password, user_input ) == 0 )
14      cout << "Correct password. Welcome to the system ..." << endl ;
15    else
16      cout << "Invalid password" << endl ;
17    return 0 ;
18  }
```

In this program the user types in a password, which is stored in user_input. Line 13 compares the user_input with the internal password "secret" held in password. The function strcmp() will return 0 if there is an exact match and a welcome message is displayed; otherwise an error message is displayed.

6.4.5   Other C-string functions

strncat(str1, str2, n)

Appends the first n characters of the C-string str2 to the C-string str1.

```
strncmp( str1, str2, n )
```
Identical to `strcmp( str1, str2 )`, except that at most, n characters are compared.

```
strncpy( str1, str2, n )
```
Copies the first n characters of `str2` into `str1`.

6.4.6　Converting numeric C-strings to numbers

Each character of the string "123" is stored in one byte of memory in the ASCII representation, as shown below.

Character:	'1'	'2'	'3'	'\0'
ASCII value in decimal:	49	50	51	0
ASCII value in binary:	00110001	00110010	00110011	00000000

This is very different from the way in which an integer value of 123 is stored. Integer values are held in binary, not ASCII format. An integer value of 123 is represented in binary as:

00000000	01111011

The functions `atoi()`, `atol()` and `atof()` convert a numeric C-string to its binary equivalent. To use any of these functions, include the preprocessor directive:

```
#include <cstdlib>
```

at the beginning of the program.
For example:

```
char str[] = "123" ;
int int_number ;
double double_number ;

int_number = atoi( str ) ; // C-string to an integer.
double_number = atof( str ) ; // C-string to a double float.
```

These functions will ignore any leading whitespace characters and stop converting when a character that cannot be part of the number is reached. For example, `atoi()` will stop when it reaches a decimal point, but `atof()` will accept a decimal point, because it can be part of a decimal number.

函数 atoi( )、atol( )和 atof( )把数值型的 C 风格字符串转化成它的二进制等价形式。要使用这些函数，必须在程序的开头包含预处理指令#include <cstdlib>。

这些函数会忽略字符间的空格或制表符，并且在遇到第一个不能作为数值部分的字符时停止转换。例如，函数 atoi( )遇到小数点时就会停止转换，但是函数 atof( )则能接受小数点，因为它可以作为小数的一部分。

6.5　C++ strings（C++ 字符串）

The C-strings discussed above can be awkward to use and tend to be error-prone. Common errors include attempting to access elements outside the array bounds, not using the function `strcpy()` to one C-string to another and not using `strcmp()` to compare two C-strings.

前面介绍的 C 风格字符串不便于使用，而且很容易出错。常见的错误有：试图访问数组范围以外的元素，没有使用函数 strcpy( )来

In addition to using C-strings, the newer and more convenient C++ strings can also be used. This doesn't mean that C-strings can be completely ignored. On the contrary, C-strings are still important because of the large quantity of software written in C++ using C-strings. The next program demonstrates C++ strings by prompting the user for a password and checking it against the correct password held in memory.

Program Example P6G

```
1   // Program Example P6G
2   // Program to demonstrate C++ strings.
3   #include <iostream>
4   #include <string>
5   using namespace std ;
6
7   int main( void )
8   {
9     string password = "secret" ;
10    string user_input ;
11    cout << "Enter Password: " ;
12    cin >> user_input ;
13    if ( password == user_input )
14      cout << "Correct password. Welcome to the system ..." << endl ;
15    else
16      cout << "Invalid password" << endl ;
17    return 0 ;
18  }
```

A sample run of this program follows:

```
Enter Password: secret
Correct password. Welcome to the system ...
```

To use C++ strings, the program must contain the line 4:

```
#include <string>
```

The `string` data type is not built into C++ like other data types such as `int`, `float` and `char`. In C++ the `string` data type is defined by a class. Although classes are not discussed until chapter 8, it is not necessary to know the details of a class in order to use it. This, in fact, is one of the strengths of a class. Suffice for the moment to say that a class introduces a new data type into a language. The new data type can then be used like any of the built-in data types.

Line 9 defines a string called `password` and initialises it to the correct password "secret". This is analogous to defining and initialising a variable of a built-in data type.

Line 10 defines another string called `user_input` with no initial value assigned.

Line 12 reads a value for `user_input` and line 13 checks if this

实现字符串之间的复制，没有使用函数strcmp( )来比较两个字符串，等等。

除了使用C风格字符串，还可以使用较新的、更加方便的C++字符串。但这并不表示可以完全忽略C风格字符串。恰恰相反，由于大量的用C++开发的软件都使用C风格字符串，因此C风格字符串仍然很重要。

与其他数据类型如整型、浮点型和字符型不一样，string数据类型不是C++语言固有的一种数据类型。在C++中，string数据类型是由类来定义的。我们将在第8章对类进行讨论，在使用一个类时没有必要去了解这个类的细节。事实上，string数据类型的功能很强大。此时知道它是语言中新引入的一种数据类型就够了。这种新的数据类型可以像其他内置数据类型一样被使用。

第12行将一个值读取到字符串

value is identical to the correct password "secret", held in the string variable password. Note that there is no need for strcmp() when comparing C++ strings as there is when comparing C-strings. C++ strings are compared using the same relational operators (==, <, >, etc) as the built-in (int, float, etc) data types.

Compare this program with the equivalent program P6F to see how much easier it is to use C++ strings rather than C-strings.

There are many useful functions associated with C++ strings, making them more powerful and convenient than C-strings. These functions are called string member functions. The following programs, which are useful for future reference, demonstrate a selection of string member functions.

user_input中，第13行核对字符串 user_input和password中 的 字 符 串是否一致。注意，在比较C++ 字符串时，不必像比较C风格字符串那样使用函数strcmp( )。像比较其他内置数据类型（int, float, 等等）一样，可以直接使用关系运算符（==, <, >, 等等）比较C++字符串。

有很多有用的函数与C++字符串相关联，这些函数使C++字符串比C风格字符串强大和方便得多。这些函数称为string类型的成员函数。

6.5.1 String initialisation and assignment

Program Example P6H

```
1   // Program Example P6H
2   // Program to demonstrate C++ string initialisation and assignment.
3   #include <iostream>
4   #include <string>
5   using namespace std ;
6
7   int main( void )
8   {
9       // String initialisation examples.
10      string str1 = "ABCDEFGHI" ; // Define a string and initialise it.
11      string str2( 11, '-' ) ;    // Define a string of 11 dashes.
12      string str3 = "This is the first part"
13                    " and this is the second part." ;
14      string str4 = str2 ; // Initialise str4 with str2.
15      string str5 ; // str5 has no initial value.
16
17      cout << "After initialisations:" << endl
18           << " str1=" << str1 << endl
19           << " str2=" << str2 << endl
20           << " str3=" << str3 << endl
21           << " str4=" << str4 << endl
22           << " str5=" << str5 << endl ;
23
24      // String assignment examples.
25      str1 = "ABCD" ;
26      str2.assign( 3, '.' ) ; // Assign 3 dots to str2.
27      cout << "After the 1st and 2nd assignments:" << endl
28           << " str1=" << str1 << endl
29           << " str2=" << str2 << endl ;
30      // Can also assign a part of another string (a sub-string).
31      // Assign 3 characters, starting at the character with index 1.
32      // The index starts at 0, so index 1 is the 2nd character.
```

```
33    str5.assign( str1, 1, 3 ) ; // Assign "BCD" to str5.
34    cout << "After the 3rd assignment:" << endl
35         << " str5=" << str5 << endl ;
36
37    // Swapping strings.
38    cout << "Before swapping str1 and str2:" << endl
39         << "  str1=" << str1 << endl
40         << "  str2=" << str2 << endl ;
41    str1.swap( str2 ) ; // swap str1 and str2.
42    cout << "After swapping str1 and str2:" << endl
43         << "  str1=" << str1 << endl
44         << "  str2=" << str2 << endl ;
45    return 0 ;
46 }
```

The output from this program is:

```
After initialisations:
  str1=ABCDEFGHI
  str2=-----------
  str3=This is the first part and this is the second part.
  str4=-----------
  str5=
After the 1st and 2nd assignments:
  str1=ABCD
  str2=...
After the 3rd assignment:
  str5=BCD
Before swapping str1 and str2:
  str1=ABCD
  str2=...
After swapping str1 and str2:
  str1=...
  str2=ABCD
```

Line 10 defines and initialises a C++ string str1.

Line 11 shows how to assign a C++ string with a number of identical characters.

As shown on lines 12 and 13, the string that is being assigned can be on two or more lines. This is useful for assigning a long string of characters to a C++ string.

正如第12行和第13行所示，可以用两行或多行来为C++字符串赋值。这种方法对于把一个很长的字符串赋值给C++字符串的情况是很有用的。

Line 14 shows how to define and initialise a C++ string with a previously defined C++ string. On line 15, str5 is defined but is not given an initial value; str5 is called an empty string.

Line 25 is a simple assignment of a character string to str1.

Line 26 uses the string member function assign() to give str2 a value of "...".

Line 33 uses assign() to assign part of str1 to str5. The values in parentheses are the string to assign from, the starting position and the

第33行使用函数assign( )把str1字符串的一部分赋值给str5。圆

number of characters to assign. The position in the C++ string starts at 0, so that the first character is in position 0, the second character is in position 1 etc.

Line 41 demonstrates the string member function swap() by swapping the two strings str1 and str2. The same can be achieved by the following:

括号中的值依次为被赋值的源字符串、赋值的起始位置、赋值的字符个数。C++字符串的起始位置为0，所以第一个字符的位置为0，第二个字符的位置为1，以此类推。

```
string temp = str1 ;
str1 = str2 ;
str2 = temp ;
```

6.5.2 String concatenation

Program Example P6I

```
1   // Program Example P6I
2   // Program to demonstrate C++ string concatenation.
3   #include <iostream>
4   #include <string>
5   using namespace std ;
6
7   int main( void )
8   {
9     string str1 = "ABCD", str2, str3 ;
10
11    str2.assign( 3, '.' ) ; // Assign 3 dots to str2.
12
13    // Concatenate str2 to str1 and assign to str3.
14    str3 = str1 + str2 ;  // With strings, + means concatenate.
15    cout << "After the 1st concatenation:" << endl
16         << " str1=" << str1 << endl
17         << " str2=" << str2 << endl
18         << " str3=" << str3 << endl ;
19
20    // Can also use += to concatenate.
21    str3 += "etc." ;   // same as str3 = str3 + "etc."
22    cout << "After the 2nd concatenation:" << endl
23         << " str3=" << str3 << endl ;
24
25    // Can also use append to concatenate.
26    str3.append ( ", etc., etc." ) ;
27    cout << "After the 3rd concatenation:" << endl
28         << " str3=" << str3 << endl ;
29
30    // Can also append a sub-string.
31    string str4 = "It is near the end of the program." ;
32    str3 = "This is " ;
33    // Append 7 characters, starting at the 12th character position.
34    str3.append( str4, 11, 7 ) ;  // Append "the end" to str3.
35    cout << "After the 4th concatenation:" << endl
36         << " str3=" << str3 << endl ;
37
```

```
38    // Finally append a repetition of a character.
39    str3.append( 3, '.' ) ;   // Append 3 dots.
40    cout << "After the 5th concatenation:" << endl
41        << " str3=" << str3 << endl ;
42    return 0 ;
43 }
```

The output from this program is:

```
After the 1st concatenation:
  str1=ABCD
  str2=...
  str3=ABCD...
After the 2nd concatenation:
  str3=ABCD...etc.
After the 3rd concatenation:
  str3=ABCD...etc., etc., etc.
After the 4th concatenation:
  str3=This is the end
After the 5th concatenation:
  str3=This is the end...
```

6.5.3   String length, string indexing and sub-strings

Program Example P6J

```
1   // Program Example P6J
2   // Program demonstrates
3   // (a) how to get the length of a C++ string
4   // (b) how to access each individual character of a C++ string
5   // (c) how to get a part (a sub-string) of a string using substr().
6   #include <iostream>
7   #include <string>
8   using namespace std ;
9
10    int main( void )
11    {
12      string str1 = "ABCDEFGH" ;
13      int len1 ;
14
15    len1 = str1.length() ;   // Store the length of str1 in len1.
16
17    // Can access each character of a string - like C-strings
18    // e.g. change the first and last characters.
19    str1[0] = '*' ;
20    str1[len1-1] = '*' ;
21    // Index start at 0 and ends at (len1-1).
22    // No index checking is done using [].
23
24    // It is much safer to check the index value to ensure it is
25    // not out of range by using the string member function at().
26    str1.at( 0 ) = 'A' ;
27    str1.at( len1 - 1 ) = 'H' ;
28
```

```
29    // Display a space between each character of str1.
30    cout << str1 << " with a space between each character:" << endl ;
31    for ( int i = 0 ; i < len1 ; i++ )
32      cout << str1.at( i ) << ' ' ;
33      cout << endl ;
34
35    // Demonstration of substr() to extract part of a C++ string.
36    // The 1st argument is a starting position and the 2nd
37    // argument is the number of characters to extract.
38    string str2 = "ABCDEFGH" ;
39    cout << "Demonstration of substr:" << endl << " " ;
40    cout << "The first four characters of " << str2<< " are "
41        << str2.substr( 0, 4 ) << endl << " "
42        << "The middle two characters of " << str2 << " are "
43        << str2.substr( 3, 2 ) << endl << " "
44        << "The last three characters of " << str2 << " are "
45        << str2.substr( 5,3 ) << endl ;
46    return 0 ;
47  }
```

The output from this program is:

```
ABCDEFGH with a space between each character:
A B C D E F G H
Demonstration of substr:
  The first four characters of ABCDEFGH are ABCD
  The middle two characters of ABCDEFGH are DE
  The last three characters of ABCDEFGH are FGH
```

6.5.4 String replace, erase, insert and empty strings

Program Example P6K

```
1   // Program Example P6K
2   // Program to demonstrate replace, erase, insert and empty.
3   #include <iostream>
4   #include <string>
5   using namespace std ;
6
7   int main( void )
8   {
9     string str1 = "ABCDE" ;
10    string str2 = "abcdefghij" ;
11
12    // Replace 3 characters from str1
13    // starting at the 2nd character position with 4 characters
14    // from str2, starting at the 3rd character position.
15    // N.B. character at position 0 is the first character.
16    str1.replace( 1, 3, str2, 2, 4 ) ;
17    cout << "After the 1st replacement:" << endl
18        << "  str1=" << str1 << endl ;
19
20    // Replace 3 characters from str1
21    // starting at the 2nd character position
```

```
22   // with all the characters from str2.
23   str1 = "ABCDE" ;
24   str1.replace( 1, 3, str2 ) ;
25   cout << "After the 2nd replacement:" << endl
26        << "  str1=" << str1 << endl ;
27
28   // Erase from the 10th character position to the end of str1.
29   str1.erase( 9 ) ;
30   cout << "After the 1st erase:" << endl
31        << "  str1=" << str1 << endl ;
32
33   // Erase 2 characters starting at the 5th character position.
34   str1.erase( 4, 2 ) ;
35   cout << "After the 2nd erase:" << endl
36        << "  str1=" << str1 << endl ;
37
38   // Erase the entire string.
39   str1.erase() ;
40   cout << "After the 3rd erase:" << endl
41        << "  str1=" << str1 << endl ;
42
43   // Are there any characters in str1?
44   if ( str1.empty() )   // empty returns TRUE or FALSE
45     cout << "str1 is empty" << endl ;
46   else
47     cout << "str1 is not empty" << endl ;
48   // Starting at the 2nd character of str2, insert 6 characters
49   // at the 5th character position of str1.
50   str1 = "ABCDEFG" ;
51   str1.insert( 4, str2, 1, 6 ) ;
52   cout << "After the 1st insert:" << endl
53        << "  str1=" << str1 << endl ;
54
55   // Insert the entire str2
56   // at the 4th character position of str1.
57   str1 = "ABCDEFG" ;
58   str1.insert( 3, str2 ) ;
59   cout << "After the 2nd insert:" << endl
60        << "  str1=" << str1 << endl ;
61   return 0 ;
62 }
```

The output from this program is:

```
After the 1st replacement:
  str1=AcdefE
After the 2nd replacement:
  str1=AabcdefghijE
After the 1st erase:
  str1=Aabcdefgh
After the 2nd erase:
  str1=Aabcfgh
```

```
After the 3rd erase:
  str1=
  str1 is empty
After the 1st insert:
  str1=ABCDbcdefgEFG
After the 2nd insert:
  str1=ABCabcdefghijDEFG
```

6.5.5 String searching

Program Example P6L

```
1  // Program Example P6L
2  // Program to demonstrate string searching.
3  #include <iostream>
4  #include <string>
5  using namespace std ;
6
7  int main( void )
8  {
9    string str1 = "ABCDEFABCDEF" ;
10   int p ;
11
12   // Find the first occurrence of "CDE" in str1.
13   p = str1.find( "CDE" ) ;
14   // The variable p holds the position of the
15   // first occurrence of "CDE" in str1.
16   // If "CDE" is not in str1, p = -1.
17   cout << "Results of 1st search:" << endl << " " ;
18   if ( p == -1 )
19     cout << "CDE Not Found in str1" << endl ;
20   else
21     cout << "First Occurrence of CDE Found at " << p << endl ;
22
23   // Reverse find - the last occurrence of "CDE"
24   p = str1.rfind( "CDE" ) ;
25   cout << "Results of 2nd search:" << endl << " " ;
26   if ( p == -1 )
27     cout << "CDE Not Found" << endl ;
28   else
29     cout << "Last Occurrence of CDE Found at " << p << endl ;
30
31   // Find the first occurrence of any one of a number of characters.
32   p = str1.find_first_of( "ED" ) ; // Find either E or D.
33   cout << "Results of 3rd search:" << endl << " " ;
34   if ( p == -1 )
35     cout << "E or D Not Found in str1" << endl ;
36   else
37     cout << "E or D First Found at " << p << endl ;
38
39   // Find the last occurrence of any one of a number of characters.
40   p = str1.find_last_of( "ED" ) ;
```

```
41    cout << "Results of 4th search:" << endl << " ";
42    if ( p == -1 )
43      cout << "E or D Not Found in str1" << endl ;
44    else
45      cout << "E or D Last Found at " << p << endl ;
46
47    // Find the first occurrence of any character that is not
48    // one of a number of characters.
49    p = str1.find_first_not_of( "ABC" ) ;
50    cout << "Results of 4th search:" << endl << " " ;
51    if ( p == -1 )
52      cout << "No Characters Other than A, B or C Found in str1"
53          << endl ;
54    else
55      cout << "A Character Other than A, B or C First Found at "
56          << p << endl ;
57
58    // Find the last occurrence of any character that is not
59    // one of a number of characters.
60    p = str1.find_last_not_of( "ABC" ) ;
61    cout << "Results of 5th search:" << endl << " " ;
62    if ( p == -1 )
63      cout << "No Characters Other Than A, B or C Found in str1"
64          << endl ;
65    else
66      cout << "A Character Other Than A, B or C Last Found at "
67          << p << endl ;
68    return 0 ;
69 }
```

The output from this program is:

```
Results of 1st search:
  First Occurrence of CDE Found at 2
Results of 2nd search:
  Last Occurrence of CDE Found at 8
Results of 3rd search:
  E or D First Found at 3
Results of 4th search:
  E or D Last Found at 10
Results of 4th search:
  A Character Other than A, B or C First Found at 3
Results of 5th search:
  A Character Other Than A, B or C Last Found at 11
```

6.5.6   String comparisons

Program Example P6M

```
1  // Program Example P6M
2  // Program to demonstrate string comparisons.
3  // Strings are compared on the basis of the ASCII codes of their
4  // individual characters.
```

```
5   #include <iostream>
6   #include <string>
7   using namespace std ;
8
9   int main( void )
10  {
11    string str1 = "ABCDEFGH" ;
12    string str2 = "BCD" ;
13    int result ;
14
15    // C++ strings can be compared with the standard comparison
16    // operators ==, !=, <=, >= < and >
17    cout << "After the standard comparison operators:"<<endl<<" ";
18    if ( str1 == str2 )
19      cout << "str1 and str2 are equal" << endl ;
20    if ( str1 < str2 ) // Does str1 come before str2?
21      cout << "str1 is less than str2" << endl ;
22    if ( str1 > str2 ) // Does str1 come after str2?
23      cout << "str1 is greater than str2" << endl ;
24
25    // The results of string comparisons can be stored in a variable
26    // using compare.
27    result = str1.compare( str2 ) ;
28    // result is < 0 if the first differing character in str1 is less
29    // than the character in the same position in str2.
30    // result is 0 if all the characters of str1 and str2
31    // are equal and the two strings are the same length.
32    // Otherwise result is > 0.
33    cout << "After the 1st compare:" << endl << " " ;
34    if ( result == 0 )
35      cout << "str1 and str2 are equal" << endl ;
36    if ( result < 0 )
37      cout << "str1 is less than str2" << endl ;
38    if ( result > 0 )
39      cout << "str1 is greater than str2" << endl ;
40
41    // Can also compare a sub-string with a string.
42    // For example, compare the 3 character sub-string of str1
43    // starting at the second character, with all characters of str2.
44    result = str1.compare( 1, 3, str2 ) ;
45    cout << "After the 2nd compare:" << endl << " " ;
46    if ( result == 0 )
47      cout << "Characters 2,3 and 4 of str1" << endl <<
48              " and all the characters of str2 are equal" << endl ;
49    if ( result < 0 )
50      cout << "Characters 2,3 and 4 of str1 are less than" << endl <<
51              " all the characters of str2" << endl ;
52    if ( result > 0 )
53      cout << "Characters 2,3 and 4 of str1 are greater than"
54              << endl << "  all the characters of str2" << endl ;
55
```

```
56    // Can also compare sub-strings.
57    // For example, compare the 2 character sub-string of str1
58    // starting at the second character,
59    // with the 2 character sub-string of str2,
60    // starting at the first character.
62    result= str1.compare( 1, 2, str2 ,0, 2 ) ;
63    cout << "After the 3rd compare:" << endl << " " ;
64    if ( result == 0 )
65      cout << "Characters 2 and 3 of str1 " << endl <<
66             "  and characters 1 and 2 of str2 are equal" << endl ;
67    if ( result < 0 )
68      cout << "Characters 2 and 3 of str1 are less than" << endl <<
69             "  characters 1 and 2 of str2" << endl ;
70    if ( result > 0 )
71      cout << "Characters 2 and 3 of str1 are greater than"
72            << endl << "  characters 1 and 2 of str2" << endl ;
73    return 0 ;
74  }
```

The output from this program is:

```
After the standard comparison operators:
  str1 is less than str2
After the 1st compare:
  str1 is less than str2
After the 2nd compare:
  Characters 2,3 and 4 of str1
  and all the characters of str2 are equal
After the 3rd compare:
  Characters 2 and 3 of str1
  and characters 1 and 2 of str2 are equal
```

6.5.7 String input

Program Example P6N

```
1   // Program Example P6N
2   // Program to demonstrate C++ string input.
3   #include <iostream>
4   #include <string>
5   using namespace std ;
6
7   int main( void )
8   {
9     string str1, str2 ;
10
11    // Demonstration of getline.
12    // getline reads all characters up to a delimiter.
13    cout << "Demonstration of getline:" << endl ;
14    cout << " Type a string and press Enter:" ;
15    getline( cin, str1, '\n' ) ;
16    // The 3rd argument is the delimiter. If omitted the delimiter
17    // is assumed to be '\n'. The above can also be written as
18    // getline( cin, str1 ) ;
```

```
19   cout << " You typed:" << str1 << endl ;
20
21   // Demonstration of input stream operator >> with strings.
22   // >> skips any leading whitespace characters,
23   // then reads input until a whitespace character is read
24   // i.e. >> reads a word.
25   cout << "Demonstration of >>:" << endl ;
26   cout << " Type a string and press Enter:" ;
27   cin >> str2 ;
28   cout << " You typed:" << str2 << endl ;
29   return 0 ;
30 }
```

A sample run of this program follows:

```
Demonstration of getline:
  Type a string and press Enter: This is a string with whitespaces
  You typed:This is a string with whitespaces·
Demonstration of >>:
  Type a string and press Enter:    This is a string with whitespaces
  You typed:This
```

cin >> str is used to input one word into the string variable str. getline(cin, str) is used to input one line from cin into the string variable str.

cin >> str用于将一个单词读取到 string 类型的变量str中。
getline(cin, str)用于从输入流cin 中将一行读取到string类型的变量str中。

6.5.8　String conversions

Program Example P60

```
1  // Program Example P60
2  // Program to demonstrate converting a C++ string to
3  // and from a C-string.
4  #include <iostream>
5  #include <string>
6  #include <cstring>
7  using namespace std ;
8
9  int main( void )
10 {
11   char c_string[6] ;
12   string cpp_string = "ABCDE" ;
13   int len ;
14
15   len = cpp_string.length() ;
16   // Convert a C++ string to a C-string.
17   cpp_string.copy( c_string, len ) ;
18   // C-strings have a '\0' at the end.
19   c_string[len] = '\0' ;
20   cout << "Results of 1st conversion:" << endl << " " ;
21   cout << "The C-string is:" << c_string << endl ;
22
23   // Convert the 2 character sub-string of cpp_string
24   // starting at the first character.
```

```
25    cpp_string.copy( c_string, 2, 0 ) ;
26    c_string[2] = '\0' ;
27    cout << "Results of 2nd conversion:" << endl << " " ;
28    cout << "The C-string is:" << c_string << endl ;
29
30    strcpy ( c_string, "abcde" ) ;
31
32    // Convert a C-string to a C++ string.
33    // A simple assignment is all that is required.
34    cout << "Results of 3rd conversion:" << endl << " " ;
35    cpp_string = c_string ;
36    cout << "The C++ string is:" << cpp_string << endl ;
37    // Note: A string literal such as "ABCDE" is in fact a C-string,
38    // so line 12 is another example of converting a C-string to a
39    // C++ string.
40    return 0 ;
41 }
```

The output from this program is:

```
Results of 1st conversion:
  The C-string is:ABCDE
Results of 2nd conversion:
  The C-string is:AB
Results of 3rd conversion:
  The C++ string is:abcde
```

6.6   Arrays of strings（string 类型的数组）

Just like any other data type, arrays of C-strings and arrays of C++ strings can be defined. Arrays of C++ strings are much easier to work with and are demonstrated in the next program.

像其他数据类型一样，也可以定义C风格字符串数组和C++字符串数组。C++字符串数组使用起来更简单一些。

Program Example P6P

```
1  // Program Example P6P
2  // Program to demonstrate an array of C++ strings.
3  #include <iostream>
4  #include <string>
5  using namespace std ;
6
7  int main( void )
8  {
9    // Define an array of twelve strings.
10   string months[12] = { "January", "February", "March",
11                         "April", "May", "June", "July",
12                         "August", "September", "October",
13                         "November", "December" } ;
14
15   // Display the months of the year.
16   cout << "The months of the year are:" << endl ;
17   for ( int i = 0 ; i < 12 ; i++ )
18     cout << months[i] << endl ;
19   return 0 ;
20 }
```

Lines 10 to 13 define and initialise an array of C++ strings. The first element of the array, `months[0]`, contains `"January"`, the second element, `months[1]`, contains `"February"`, and so forth. The loop in lines 17 to 18 displays each element of the array.

6.7　**Character classification**（字符分类）

There are a number of C++ functions that can be used to test the value of a single character. These functions return a true (non-zero integer) value or a false (zero integer) value depending on whether or not the character belongs to a particular set of characters.

有很多C++函数可以用来测试单个字符的值。根据这个字符是否属于一个特定的字符集，这些函数返回值为真（非零整数）或者为假（整数0）。

Function	Character set
isalnum	Alphanumeric character: A-Z, a-z, 0-9
isalpha	Alphabetic character: A-Z, a-z
isascii	ASCII character: ASCII codes 0-127
iscntrl	Control character: ASCII codes 0-31 or 127
isdigit	Decimal digit: 0-9
isgraph	Any printable character other than a space
islower	Lowercase letter: a-z
isprint	Any printable character, including a space
ispunct	Any punctuation character
isspace	Whitespace character: \t,\v,\f,\r,\n or space
	ASCII codes 9-13 or 32
isupper	Uppercase letter: A-Z
isxdigit	Hexadecimal digit: 0-9 and A-F

For example:

```cpp
char ch ;
cin >> ch ;

if ( isupper( ch ) )
  cout<< ch << " is an uppercase character" << endl ;
```

C++ has also two further functions that are used to covert the case of a character: `tolower` and `toupper`.

C++还有两个用于转换字符大小写状态的函数：tolower和toupper。

Function	Character set
tolower	Converts an uppercase character to lowercase.
toupper	Converts a lowercase character to uppercase.

For example:

```cpp
char ch = 'a' ;
ch = toupper( ch ) ;
cout << "Convert to uppercase " << ch << endl ;
ch = tolower( ch ) ;
cout << "Back to lowercase " << ch << endl ;
```

As a practical example, the following program inputs a name and displays the name with the first letter of the forename and surname capitalised. The program assumes that the forename and surname are separated by at least one space.

Program Example P6Q

```
1   // Program Example P6Q
2   // Program to input a name and capitalise the first alphabetical
3   // character of the forename and surname. The program ignores
4   // non-alphabetical characters at the start of the forename
5   // and assumes at least one space between the forename and surname.
6   #include <iostream>
7   #include <string>
8   using namespace std ;
9
10  int main( void )
11  {
12    string in_name ;
13
14    cout << "Type a forename and surname and press Enter: " ;
15    getline( cin, in_name ) ;
16
17    // Ignore all characters until the first alphabetical character
18    // of the forename is reached.
19    int i=0 ;
20    while( !isalpha( in_name[i] ) )
21      i++ ;
22
23    // Capitalise the first character of the forename.
24    in_name[i] = toupper( in_name[i] ) ;
25
26    // Ignore all characters in forename until a space is reached.
27    while( !isspace( in_name[i] ) )
28      i++ ;
29
30    // Ignore all characters until the first letter of the surname
31    // is reached.
32    while( !isalpha( in_name[i] ) )
33      i++ ;
34
35    // Capitalise the first letter of the surname.
36    in_name[i] = toupper( in_name[i] ) ;
37
38    cout << "Formatted Name: " << in_name << endl ;
39    return 0 ;
40  }
```

A sample run of this program is:

Type a forename and surname and press Enter: john smith

```
Formatted Name: John Smith
```

Programming pitfalls

1. Double quotation marks (") are used for string literals; single quotation marks (') are used for character constants. For example, `"abcd"` and `'e'`.

2. When you allocate space for a C-string you must allow for the terminating null character `'\0'`. For example, to store a C-string of twenty characters you must define a character array with twenty-one elements.

3. When using `strcpy( str1, str2 )` to copy C-strings, the string `str2` is copied to the string `str1`, and not the other way around.

4. To compare two C-strings use `strcmp()`. For example, if `str1` and `str2` are C-strings, then the statement

```
if ( str1 == str2 ) // OK for C++ strings,
                     // but not for C-strings.
```

is incorrect. The correct statement is

```
if ( strcmp(str1, str2 ) == 0 )
```

5. Accessing a character in a C or C++ string that is outside the subscript range of the string using `[]` will result in a runtime error. For example,

```
char   cs[5] = "abcd" ; // a 4 character C-string
string cpps = "abcd" ; // a 4 character C++ string
```

`cs[5]`, `cs[6]`, `cs[7]` etc are invalid, as are `cpps[5]`, `cpps[6]`, `cpps[7]` etc.

With a C++ string this error can be detected using the `string` member function `at()`.

1. 双引号（ `"` ）用于表示字符串，单引号（ `'` ）用于表示字符，例如"abcd"和'e'。

2. 定义C风格字符串时，一定要给'\0'留出空间。例如，要存储包含20个字符的一个C风格字符串，必须定义有21个元素的数组。

3. 函数strcpy(str1, str2)是把字符串str2 复制到str1 中的，不要弄反了。

4. 在比较两个C风格字符串时必须使用函数strcmp( )。

5. 以数组下标方式访问超出C或C++字符串长度范围的字符将导致一个运行时错误。使用C++ 类成员函数at( )可以检测该错误。

Quick syntax reference

	Syntax	Examples

To define a C-string	`char variable_name[n+1] ;`
	`// n is the number of characters`
	`// in the string.`
		`char c_str[11] ;`
		`// c_str can hold ten`
		`// characters plus the`
		`// null character '\0'.`

C-string input	`cin >> variable_name ;`
	`cin.getline( variable_name,n+1,delim ) ;`
	`// n is the number of characters`
	`// in the string.`
	`// delim is the delimiter,`
	`// default value of delim is '\n'.`
		`cin >> c_str ;`
		`cin.getline( c_str, 11 ) ;`

C-string length	`size = strlen( c_string ) ;`	`len = strlen( c_str ) ;`

(cont.)

	Syntax	Examples
Comparing C-strings	`strcmp( c_string1, c_string2 )`	`if ( strcmp(c_str1,c_str2)` `== 0 )` `cout << "Identical" ;`
Copying a C-string	`strcpy( destination, source ) ;`	`// Copy c_str2 to c_str1.` `strcpy( c_str1, c_str2 ) ;`
To define a C++ string	`string variable_name ;`	`string cpp_str ;`
C++ string input	`cin >> variable_name ;` `getline( cin, variable_name, delim ) ;` `// delim is the delimiter,` `// default value of delim is '\n'.`	`cin >> cpp_str ;` `getline( cin, cpp_str ) ;`
C++ string length	`size = variable_name.length() ;`	`len = cpp_str.length() ;`

Exercises

1. What is the output from the following program segment?

```
char str1[] = "abc" ;
char str2[] = "ABCD" ;
cout << str1 << endl << strlen( str1 ) << endl ;
if ( strcmp( str1, str2 ) == 0 )
  cout << str1 << "==" << str2 << endl ;
else
  if ( strcmp( str1, str2 ) < 0 )
    cout << str1 << "<" << str2 << endl ;
else
  if ( strcmp( str1, str2 ) > 0 )
    cout << str1 << ">" << str2 << endl ;
char str3[8] ;
strcpy( str3, str1 ) ;
strcat ( str3, str2 ) ;
cout << str3 << endl << strlen( str3 ) << endl ;
str3[6] = 'x' ;
cout << str3 << endl ;
```

2. Modify exercise 1 to use C++ strings rather than C-strings.

3. Given the following:

```
char c_str1[18] ;
char c_str2[6] = "abcde" ;
```

what is in `c_str1` after each of the following?

(a) `strcpy( c_str1, "A string" ) ;`

(b) `strcat ( c_str1, " of text." ) ;`

(c) `strncpy( c_str1, c_str2, 1 ) ;`

4. Write a program to input a C-string from the keyboard and replace each space in the string with the character '_'.

5. What is the output from the following program segment?

```
string str = "ABCDEFGHIJ" ;
cout << str << endl << str.length() << endl ;
str.replace( 4, 2, "123456" ) ;
str.at( 3 ) = '0' ;
cout << str << endl << str.length() << endl ;
str.erase( 10, 2 ) ;
cout << str << endl << str.length() << endl ;
cout << str.substr( 3, 7 ) << endl ;
str += "KLMN" ;
cout << str << endl << str.length() << endl ;
str.insert( 10, "7890" ) ;
cout << str << endl << str.length() << endl ;
cout << str.find( "0" ) << endl ;
```

6. Write a program to read in three names from the keyboard and display them in alphabetical order.

7. Write a program to read in a line of text from the keyboard and calculate the average length of the words in that line. Assume each word in the line is separated from the next by at least one space. Allow for punctuation marks. Use C++ strings.

8. Modify exercise 7 to display the number of words in the line with lengths of

   (a) 1

   (b) 2 to 5

   (c) 6 to 10

   (d) 11 to 20

   (e) 21 and above.

9. Write a program to input a user's first and last name and generate a password for the user.
   The password is made up from the first, middle and last character from the first and last name.
   Example:

   Enter your first name: Yang

   Enter your second name: Liwei

   Your password is YngLwi

10. Write a program to ask a user for their name. The user's name is then compared with a list of names held in an array in memory. If the user's name is in this list, display a suitable greeting; otherwise display the message "Name not found".

11. Write a program to input a string. If every character in the string is a digit ('0' to '9'), then convert the string to an integer, add 1 to it, and display the result. If any one of the characters in the string is not a digit, display an error message.

12. The following is a list of countries and their capital cities.

Australia	Canberra
Belgium	Brussels
Denmark	Copenhagen
United Kingdom	London
France	Paris
Greece	Athens
Ireland	Dublin

Write a program to input a country and display the capital city of that country.

13. Initialise an array of strings with the following quotations:

   "There is no reason for any individual to have a computer in their home."

   "Computers are useless. They can only give you answers."

   "To err is human, but to really foul things up requires a computer."

   "The electronic computer is to individual privacy what the machine gun was to the horse cavalry."

   Input a word from the keyboard and display all quotations, if any, containing that word.

14. Input two strings from the keyboard and check if they are anagrams of each other.

   Hint: Take each character of the first string and check that it exists in the second string. If it doesn't then the strings are not anagrams of each other. If the character does exist in the second string, remove it and continue to the next character of the first string. When all the characters of the first string have been processed, the second string should be empty. If it is, then the two strings are anagrams of each other. Use C++ strings.

15. Write a program to input a string and check whether or not it is a valid e-mail address.

   A valid e-mail address

      cannot start or end with @

      cannot start or end with .

      must contain @ only once

      must contain . at least once after @

Chapter Seven
Functions
第 7 章　函　　数

7.1　Introduction（引言）

The programs used in previous chapters are relatively small programs used to demonstrate a particular topic in C++. Typical commercial programs have hundreds or even thousands of lines of code. In order to reduce the complexity involved in writing such large programs, they have to be broken into smaller, less complex parts. Functions, along with classes that are covered in chapter 8, enable the programmer to do this. Functions and classes are the building blocks of a C++ program.

A function is a block of statements called by name to carry out a specific task, such as displaying headings at the top of every page of a report, reading a file, or performing a series of calculations.

C++ has a variety of built-in, pre-written, functions in the standard library. The next program demonstrates one such function, sqrt(), that calculates the square root of a number. By convention, when referring to a function the name of the function is followed by parentheses.

前面章节中的程序都是相对较小规模的程序，用来演示C++的特定专题。典型的商用程序通常有几百甚至上千行代码。为了降低编写如此大规模代码的复杂度，程序员必须将其分解为较小、较简单的模块，例如函数和第 8 章将要介绍的类。函数和类是构造C++程序的基本模块。

函数是用函数名来调用执行的具有特定功能的语句块，例如在报告的每一页的顶部显示标题、读文件，或者进行一系列的计算。

在C++的标准库中有很多固有的、预先定义的库函数。下面的程序就是使用这样的一个库函数sqrt( )来计算一个数的平方根的例子。按惯例，在引用函数名的后面紧跟着一对圆括号。

Program Example P7A

```cpp
1   // Program Example P7A
2   // Program to demonstrate the built-in function sqrt().
3   #include <iostream>
4   #include <string>
5   #include <cmath>
6   using namespace std ;
7
8   int main( void )
9   {
10    for ( int n = 1 ; n < 11 ; n ++ )
11      cout << sqrt( n ) << endl ;
12    return 0 ;
13  }
```

Line 11 *calls* the function sqrt() to calculate the square root of the value in the variable n, which successively is assigned the values 1, 2, 3, ..., 10 by the for statement on line 10.

Line 5 is required to use any of the mathematical functions. The mathematical functions have been adopted from the C programming language, hence the c in <cmath>.

Although the standard library is extensive, it does not include every function that a programmer may require. C++ allows a programmer to write functions to add to those already at hand in the standard library. The next program displays a string of text surrounded by a box of asterisks. The program includes a programmer-defined function to display a line of asterisks.

虽然标准库提供了大量标准函数，但是并非程序员所需的每个函数都包含在内。C++ 允许程序员自己编写函数加入手头已存在的标准库中。

Program Example P7B

```
1   // Program Example P7B
2   // Demonstration of a programmer-defined function.
3   #include <iostream>
4   #include <string>
5   using namespace std ;
6
7   void stars( void ) ;         Function declaration or Prototype.
8
9   int main( void )
10  {
11    string text = "some text" ;
12
13    stars() ;      // Call the function to display the top of the box.
14    cout << endl ;
15    cout << '*' ;            // Left side of the box.
16    cout << text ;           // Text in middle of the box.
17    cout << '*' << endl ;    // Right side of the box.
18    stars() ;                // Bottom of the box.
19    cout << endl ;
20    return 0 ;
21  }       ←── This is the end of main()
22
23  void stars( void )
24  {                                            Function
25    for ( int counter = 0 ; counter < 11 ; counter++ )   definition.
26      cout << '*' ;
27  }
```

The output from this program is:

```
***********
*some text*
***********
```

Like variables, functions must be declared before they are used. Line 7 declares stars to be a function. The first void on line 7 declares the type of the function stars(). Functions with type void do not return a value to the calling program. Some functions do return a value. For example the sqrt() function returns a numeric value, i.e. the square root of a number.

和变量一样，函数在使用之前也必须进行声明。第7行将stars声明为函数。前面的void是对函数stars( )类型的声明。将函数类型声明为void，表示函数不会向调用程序返回值。而某些函数确实

The void in the parentheses on line 7 informs the compiler that the function stars will not receive any data from the calling program. Some functions do receive data when called. For example, with sqrt() a number or a numeric variable is placed in the parentheses.

In summary,

void stars(void);

↑　　　　　　↑

Returns nothing.　*Receives nothing.*

The function declaration on line 7 is also called the *prototype* of the function stars().

Lines 13 and 18 call the function stars(), resulting in the top and bottom lines of stars being displayed.

Lines 23 to 27 define the function. Line 23 is called the *function header*, the remainder of the function enclosed in the braces { and } is called the *function body*.

In both lines 7 and 23 the void in the parentheses is optional. The function prototype on line 7 can therefore be written as:

```
void stars() ;
```

and the function header on line 23 can also be written as:

```
void stars()
```

需要返回一个值，例如函数sqrt( )就返回一个数值，即一个数的平方根。

第7行圆括号中的void告诉编译器stars函数将不会接收来自调用程序的任何数据。而某些函数在调用时确实是需要接收数据的，例如函数sqrt( )就要求把一个数或者数值变量放在函数名后面的圆括号中。

第7行的函数声明称为函数stars( )的函数原型。

第13行和第18行调用函数stars( )，是为了显示最上方和最下方的一行星号。

第23至27行是函数定义。第23行称为函数头，包含在一对花括号内的其余部分称为函数体。

7.2　Function arguments（函数实参）

The function stars(), as written in P7B, displays eleven asterisks every time it is called. It would be useful to have a function to display a variable number of asterisks, not just eleven. This new function could be used for displaying boxes of different sizes.

The next program modifies the function stars() to take a value passed to it and to display the number of asterisks specified in that value.

Program Example P7C

```
1   // Program Example P7C
2   // Demonstration of function arguments.
3   #include <iostream>
4   #include <string>
5   using namespace std ;
6
7   void stars( int ) ; // Function prototype.
8
9   int main( void )
10  {
11    string text = "some text" ;
```

```
                                   ┌──────→ 11 is passed to num ─────────────┐
12                                 │                                         │
13    stars( 11 ) ;       // Call the function to print 11 *s - top of box.  │
14    cout << endl ;                                                         │
15    stars( 1 ) ;        // Left side of box - 1 * only.                    │
16    cout << text ;       // Display text in the middle of the box.         │
17    stars( 1 ) ;        // Right side of box - 1 * only.                   │
18    cout << endl ;                                                         │
19    stars( 11 ) ;       // Bottom of the box - 11 *s.                      │
20    cout << endl ;                                                         │
21    return 0 ;                                                             │
22  }                                                                        │
23                        │                                                  │
24  void stars( int num ) ◄──────────────────────────────────────────────── ┘
25  {
26    for ( int counter = 0 ; counter < num ; counter++ )
27      cout << '*' ;
28  }
```

Calls to the function stars() in lines 13, 15, 17 and 19 now have a number between the parentheses (and). This number is called an *argument* and is received by the *parameter* num declared as an integer in line 24.

The parameters of a function are known only within the function. Therefore, variables with the same name can be used in main() or in any other function without a conflict occurring.

Line 26 of the function stars() now uses the variable num to decide how many times * is displayed.

Since the function is now receiving an integer value when it is called from main(), the function prototype on line 7 now has int in the parentheses.

An even more useful function than stars() would be one where you could specify not only the number but also the character to display. The next program includes such a function: disp_chars(). The program displays

```
+++++++++++
+some text+
+++++++++++
```

in the middle of the screen.

现在，在第13、15、17 和 19 行调用函数stars( )时，圆括号内有一个数值。这个数值称为实际参数（简称实参），它从第24行声明为整型的形式参数（简称形参）中接收数据。

函数的形参仅在函数内部有效，因此，在函数main( )或其他任何函数中使用同名变量不会发生冲突。

因为该函数被main( )调用时，需要从main( )接收一个整型数值，所以第 7 行的函数原型就需要在圆括号内写上int。

Program Example P7D

```
1   // Program Example P7D
2   // Demonstration of a function with two parameters.
3   #include <iostream>
4   #include <string>
5   using namespace std ;
6
7   void disp_chars( int num, char ch ) ;
```

Function interface.

```
8   // Purpose   : To display any number of any character.
9   // Parameters: The number of times to display a character and
10  //             the character to display.
11
12  int main( void )
13  {
14    string text = "some text" ;
15                          35 is passed to num and a space is passed to ch
16    // Top of box.
17    disp_chars( 35, ' ' ) ; // Display 35 spaces
18    disp_chars( 11, '+' ) ; // and eleven +s.
19    cout << endl ;
20    // Left side of box.
21    disp_chars( 35, ' ' ) ; // Display 35 spaces
22    disp_chars( 1, '+' ) ;  // and a +.
23    // Display text.
24    cout << text ;
25    // Right side of box.
26    disp_chars( 1, '+' ) ; // Display a +.
27    cout << endl ;
28    // Bottom of the box.
29    disp_chars( 35, ' ' ) ; // Display 35 spaces
30    disp_chars( 11, '+' ) ; // and eleven +s.
31    cout << endl ;
32    return 0 ;
33  }
34
35  void disp_chars( int num, char ch )
36  {
37    for ( int counter = 0 ; counter < num ; counter++ )
38      cout << ch ;
39  }
```

The function disp_chars() uses two parameters: num (the number of times to display a character) and ch (the character to display). Different values are passed to these parameters when the function is called in lines 17, 18, 21, 22, 26, 29, and 30.

The function prototype on line 7 informs the compiler that the function disp_chars will receive an int and a char from the calling program. The compiler checks that the type and number of arguments used in the function calls match the function prototype.

```
void disp_chars( int num, char ch ) ;
```

Returns nothing.　*Receives an integer and a character.*

The variable names num and ch are included in the function prototype on line 7. This is optional but can help in describing the function parameters.

The variable names used in the prototype can be any valid variable

函数原型中使用的变量名可以是任何有效的变量名，无须定义。实际应用中，这些名字通常都与函数头中定义的形参的名字相

names and need not be defined. In practice, the names are often the same as those used for the parameters in the function header. This means that the prototype (line 7) is often the same as the function header (line 35), but the prototype ends with a semicolon.

It is good practice to place comments after the function prototype to describe the function and its parameters. The prototype and the comments should provide another programmer with enough information on how to use the function. The prototype and the accompanying comments are known as the *function interface*.

同。这说明函数原型（第7行）通常与函数头（第35行）是相同的，唯一区别是函数原型的末尾多了一个分号。

在函数原型的后面写上一段注释来描述函数及其形参，是一个非常好的习惯。函数原型和注释必须给其他程序员提供足够的信息，让其了解如何使用该函数。函数原型及其随后的注释称为函数接口。

7.3 Default parameter values（默认的形参值）

Some or all of the arguments in a function call can be omitted, provided a default value is assigned to the corresponding parameter of each missing argument.

A function parameter that is not passed a value can be assigned a *default* value. This default value is specified in the prototype of the function.

For example, to give default values to the parameters of disp_chars() in program P7D, the prototype on line 7 is modified to

```
void disp_chars( int num = 1, char ch = ' ' ) ;
```

The prototype now assigns a default value of 1 to the first parameter and a space to the second parameter.

For example, the statement

```
disp_chars( 35 ) ; // The second argument is omitted.
```

is equivalent to

```
disp_chars( 35, ' ' ) ;
```

Similarly, the statement

```
disp_chars() ;      // Both arguments are omitted.
```

is equivalent to

```
disp_chars( 1, ' ' ) ;
```

If a parameter is provided with a default value, all the parameters to its right must also have a default value. This means that it is illegal to have a parameter with a default value followed by a parameter without a default value. For example,

```
void disp_chars( int num = 1, char ch ) ; // Illegal.
void disp_chars( int num, char ch = ' ' ) ; // Legal.
```

在函数调用时，函数的部分或者全部实参省略不写的前提条件是，事先已将一个默认值赋值给被省略的实参所对应的形参。没有向其传递实参值的函数形参可以被赋值一个默认的值。这个默认值需要在函数原型中指定。

如果一个函数形参被指定了默认值，那么位于它右侧的所有形参都必须指定默认值，也就是说，在一个指定了默认值的形参后面跟随没有默认值的形参是不合法的。

7.4　Returning a value from a function
　　（从函数返回一个值）

To demonstrate the returning of a value from a function, the next program calls a function with two integer values as arguments and returns the minimum of the two values.

Program Example P7E

```
1   // Program Example P7E
2   // Demonstration of the return statement.
3   #include <iostream>
4   using namespace std ;
5
6   int minimum( int num1, int num2 ) ;
7   // Purpose : returns the minimum of two integers.
8   // Parameters: two integer values num1 and num2.
9   // Returns : the minimum of num1 and num2.
10
11  int main( void )
12  {
13    int val1, val2, min_val ;
14
15    // Read in two integer values from the keyboard.
16    cout << "Please enter two integers: " ;
17    cin >> val1 >> val2 ;
18    // Find the minimum of these two values.
19    min_val = minimum( val1, val2 ) ;
20
21    cout << "Minimum of " << val1 << " and " << val2
22         << " is " << min_val << endl ;
23    return 0 ;
24  }
25
26  int minimum( int num1, int num2 )
27  {
28    if ( num1 < num2 )
29      return num1 ;
30    else
31      return num2 ;
32  }
```

Return the value of num1 *or* num1 *to* min_val

A sample run of this program is:

```
Please enter two integers: 1  2
minimum of 1 and 2 is 1
```

Because `minimum()` returns an integer value, you must tell the compiler this in two places in the program. The first place is in the function prototype on line 6, and the second place is in the function header on line 26.

必须在程序的两个地方声明 minimum( ) 的返回值为整型：第一处是第 6 行的函数原型，第二处是第 26 行的函数头。

```
int minimum( int num1, int num2 )
```

Returns an integer.   *Receives two integers.*

The `return` statement in lines 29 and 31 does two things: it terminates the function and returns the value of either `num1` or `num2` to the integer variable `min_val` in line 19.

第29行和第31行的return语句完成两件事情：结束函数的执行；将num1或num2中的一个值作为返回值，赋值给第19行的变量min_val。

The general format of the `return` statement is:

```
return expression ;
```

Examples:

```
return 10.3 ;            // Return a constant value.
return ;                 // No return value, just exit the function.
return variable ;        // Return the value of a variable.
return variable + 1 ;    // Return the value of an expression.
```

Lines 28 to 31 of `minimum()` could also be written as

```
return ( num1 < num2 ) ? num1 : num2
```

A function call can be used anywhere in a program where a variable can be used. For example, in the program P7E the variable `min_val` in line 22 may be replaced with the function call, as in:

在程序中可以引用变量的任何地方都可以进行函数调用。

```
cout << "Minimum of " << val1 << " and " << val2
     << " is " << minimum( val1, val2 ) << endl ;
```

7.5   Inline functions（内联函数）

Calling a function, passing argument values to a function and returning a value from a function all involve some processing overheads. For a large function that contains a lot of statements, the overheads involved are small in comparison with the processing done by the function itself. However, for small functions, like `minimum()` in program P7E, the overheads are relatively large in comparison with the small number of statements in the function. In this case it would be advantageous to make `minimum()` an `inline` function.

函数调用、向函数传递实参值、从函数返回一个值，都需要一些执行时间开销。对于一个规模较大的函数，相对于函数自身的执行时间开销而言，这些开销还是很小的。但是对于一个规模较小的函数，例如程序P7E 中的minimum( )，相对于执行函数内的少量语句所需的开销而言，这些开销就相对较大了。在这种情况下，把函数minimum( )变成内联函数将是非常有利的。

To make a function `inline`, precede the function prototype with the keyword `inline`. When an `inline` function is used in a program, the compiler replaces every call to the function with the program statements in the function, thereby eliminating the function overheads.

In general, consider inlining a function when the function contains one to three lines of code.

在函数原型前面加上关键字inline，即可把函数变成内联函数。当程序使用内联函数时，编译器将每个函数调用都用函数内的语句代替，这样就省去了函数调用的开销。

To make `minimum()` an `inline` function, change line 6 in program P7E to

```
inline int minimum( int num1, int num2 ) ;
```

No other changes to the program are necessary.

一般情况下，当函数内仅有1至3行代码时应考虑函数内联。

7.6 Passing arguments by value（按值传递实参）

In the functions of the previous programs, a *copy* of the argument values is passed to the function parameters. This is known as *passing by value*.

在前面程序的函数中，将实参值的一个副本传递给函数的形参，称为按值传递实参。

As only a copy of an argument is sent to the function, the value of the argument cannot be changed within the function.

Program Example P7F

```
1   // Program Example P7F
2   // Demonstration of passing arguments by value.
3   #include <iostream>
4   using namespace std ;
5
6   void any_function( int p ) ;
7
8   int main( void )
9   {
10    int a = 1 ;
11    cout << "a is " << a << endl ;
12
13    any_function( a ) ;
14
15    cout << "a is still " << a << endl;
16    return 0;
17  }
18
19  void any_function( int p )
20  {
21    cout << "p is " << p << endl ;
22    p = 2 ;
23  }
```

A copy of the value of a is passed to p

Changing the value of the parameter p *has no effect on the argument* a.

When you run this program you will get the following result:

```
a is 1
p is 1
a is still 1
```

This program changes the value of the parameter p in line 22 without having any effect on the argument a. Line 15 displays the value of the argument a, showing it to be unchanged.

The value of the parameter can be prevented from change within a function by making it a constant. To do this, place the keyword const before the parameter in lines 6 and 19.

将形参声明为常量，可以防止形参值在函数内被修改，只要将关键字const放在形参前即可。

```
6 void any_function( const int p ) ;
...
19 void any_function( const int p )
```

In this case, trying to change the value of parameter p on line 22 will cause the compiler to display an error message.

7.7 Passing arguments by reference
（按引用传递实参）

A reference is a synonym or an alias for an existing variable. For example, if a variable n is defined as:

```
int n = 1 ;
```

a reference variable r for n can be defined by the statement:

```
int& r = n ; // r is a reference to n. r is any valid identifier.
```

引用是变量的同义词或者别名。

A reference to a variable is defined by adding & after the variable's data type.

After this definition n and r both refer to the same value, as if they were the same variable. The contents of n can be accessed by either n or r.

Note that r is not a copy of n, but is merely another name for n. Both n and r refer to the same storage location.

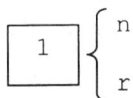

变量的引用是通过在变量数据类型后面加上&来定义的。

这样定义变量n和r以后，它们指的就都是同一个值了，就好像它们原本是一个变量一样。n的内容既可以用n来访问，也可以用r来访问。

注意，r不是n的副本，只是n的另一个名字而已。n和r指的都是同一个存储单元。

$$1 \begin{cases} n \\ r \end{cases}$$

A change to n will also result in a change to r and vice versa. For example,

对n的修改也会导致r值的改变，反之亦然。

```
n = 2 ; // Changes both n and r.
```

will result in both n and r having the value 2.

A reference must always be initialised when it is defined, since it doesn't make sense to have a reference to nothing.

定义引用变量时，必须对其进行初始化，因为定义一个不指向任何变量的引用是没有意义的。

```
int& r ; // Illegal: a reference must be initialised.
```

References are commonly used as function parameters. As an example, the next program modifies program P7F to use a reference as a function parameter.

引用通常被用作函数的形参。

Program Example P7G

```
1  // Program Example P7G
2  // Program to demonstrate passing an argument by reference.
3  #include <iostream>
4  using namespace std ;
5
6  void any_function( int& p ) ;
7
8  int main( void )
```

```
9  {
10    int a = 1 ;
11    cout << "a is " << a << endl ;
12
13    any_function( a ) ;
14
15    cout << "a is now " << a << endl ;
16    return 0 ;
17  }
18
19  void any_function( int& p )
20  {
21    cout << "p is " << p << endl ;
22    p = 2 ;
23  }
```

a and p *refer to the same storage location.*

Changing the value of the parameter p *also changes the value of the argument* a.

The output from this program is:

```
a is 1
p is 1
a is now 2
```

Line 19 now declares the parameter p on line 19 to be a reference and the function prototype on line 6 is also changed accordingly. The argument a on line 13 is now *passed by reference* to the parameter p on line 19.

Line 22 changes the value of the parameter p and hence the value of the argument a also changes. Line 15 displays the value of a, showing that it has changed from 1 to 2.

The next program is a further demonstration of passing arguments by reference. This program has a function swap_vals() that has two reference parameters. This function swaps its two parameter values around and in so doing swaps the argument values around.

第19行将形参p声明为引用变量，在第6行的函数原型中也相应地进行了修改。在第13行中，实参a按引用传递给第19行的形参p。

Program Example P7H

```
1   // Program Example P7H
2   // Demonstration of passing two arguments by reference.
3   #include <iostream>
4   using namespace std ;
5
6   void swap_vals( float& val1, float& val2 ) ;
7   // Purpose : To swap the values of two float variables.
8   // Parameters: References to the two float variables.
9
10  int main( void )
11  {
12    float num1, num2 ;
13
14    cout << "Please enter two numbers: " ;
15    cin >> num1 ;
```

```
16    cin >> num2 ;
17    // Swap values around so that the smallest is in num1.
18    if ( num1 > num2 )
19      swap_vals( num1, num2 ) ;
20    cout << "The numbers in order are "
21        << num1 << " and " << num2 << endl ;
22    return 0 ;
23 }
24
25 void swap_vals( float& val1, float& val2 )
26 {
27    float temp = val1 ;
28
29    val1 = val2 ;
30    val2 = temp ;
31 }
```

Arguments passed by reference.

A sample run of this program is:

```
Please enter two numbers: 12.1   6.4
The numbers in order are 6.4 and 12.1
```

No value is returned from swap_vals(), so line 6 declares the function as type void.

Line 25 declares the parameters val1 and val2 as references to the arguments num1 and num2 on line 19.

The statement

```
float temp = val1 ;
```

on line 27 stores the value of val1 (which is a reference to num1) in the variable temp (temp is now 12.1).

The statement

```
val1 = val2;
```

on line 29 is equivalent to

```
num1 = num2 ;
```

because val1 is a reference to num1 and val2 is a reference to num2. Therefore, num1 gets the value 6.4.

Finally, the statement

```
val2 = temp ;
```

on line 30 assigns the value of temp (12.1) to val2, and in doing so also assigns the same value to num2.

The net result is that the values in num1 and num2 are swapped around. The variable temp defined on line 27 is a *local variable* to the function swap_vals(). Local variables are known only within the function where they are defined. If a variable of the same name is defined in main() or in another function, C++ regards the two variables as different and not related in any way.

局部变量仅在定义它的函数内部才能使用。如果在函数main( )或另一个函数中定义了一个同名变量，则C++将其视为两个不同的变量，它们之间没有任何关系。

7.8 Passing a one-dimensional array to a function
（向函数传递一维数组）

To avoid the overhead of copying all the elements of an array, that passing by value would entail, arrays can only be passed by reference to a function.

The next program contains a function `sum_array()` that sums the elements of an integer array passed to it from `main()`.

为了避免按值传参引起的复制数组所有元素所需的开销，可以使用按引用传参的方法将数组传递给函数。

Program Example P7I

```
1  // Program Example P7I
2  // Demonstration of a one-dimensional array as a function argument.
3  #include <iostream>
4  using namespace std ;
5
6  int sum_array( const int array[], int no_of_elements ) ;
7  // Purpose : Sums the elements of a 1-D integer array.
8  // Parameters: An array and the number of elements in the array.
9  // Returns : The sum of the array elements.
10
11 int main( void )
12 {
13   int values[10] = { 12, 4, 5, 3, 4, 0, 1, 8, 2, 3 } ;
14   int sum ;
15
16   sum = sum_array( values, 10 ) ;
17   cout << "The sum of the array elements is " << sum << endl ;
18   return 0 ;
19 }
20
21 int sum_array( const int array[], int no_of_elements )
22 {
23   int total = 0 ;
24
25   for ( int index = 0 ; index < no_of_elements ; index++ )
26     total += array[index] ;
27   return total ;
28 }
```

Running this program will display the following:

```
The sum of the array elements is 42
```

Line 16 calls `sum_array()` to calculate the sum of the values in the array `values`, placing the result of the calculation in the variable `sum`. The arguments are the name of the array and the number of elements in the array.

Line 21 declares the function parameters `array` and `no_of_elements`. Although there is no `&` after the data type, `array` is a reference to `values`. Because arrays can only be passed by reference, `&` is not required.

The square brackets [and] are necessary to indicate that the parameter is a reference to an array. For a one-dimensional array, the number of elements is not required in the brackets. This allows the same function to be used to process arrays of different sizes without the function being modified.

The loop in lines 25 and 26 calculates the sum of the array elements and line 27 returns the result to `sum` on line 16.

Because an array argument can only be passed by reference, the values of the elements in the array can be changed from within the function, but the keyword `const` in lines 6 and 21 prevents this from happening. The `const` keyword informs the compiler that within the function `sum_array()`, `array` is read-only and cannot be modified.

The next program is a further demonstration of passing an array to a function. The array is modified in the function and hence `const` is not used for the array parameter.

The program also demonstrates an alternative way of commenting a function. Instead of comments listing the parameters and return value, pre and post condition comments are used.

In the pre condition comment, the conditions that must exist before a function may be successfully called are listed. In the post condition comment, the conditions that will exist after the function is called are listed. Whichever style is chosen, it is important that each function in a program is commented. Good comments enable a function to be easily used and understood by other programmers.

方括号[和]是必需的，它表示形参是一个数组的引用。对于一维数组，不必在方括号内指出元素的个数，这样使得同一个函数可以用于处理不同大小的数组，而无须修改程序。

因为只能按引用传递数组实参，所以数组元素的值可以在函数中被修改，但是第6行和第21行的关键字const可用于避免数组元素被意外修改。关键字const通知编译器，在函数sum_array( )内部，数组array是只读的，不能被修改。如果要在函数中修改数组元素的值，那么就不要使用const来定义形参数组。

除了列出形参和返回值这种函数注释方式，还可以使用前置和后置条件的函数注释方式。在前置条件注释中，列出成功调用函数所必需的前提条件。在后置条件注释中，列出调用函数后产生的结果。无论使用哪一种风格的注释都是可以的，重要的是在程序中务必要给每个函数都加上注释。对程序员而言，良好的注释可以使函数易读、易懂。

Program Example P7J

```
1   // Program Example P7J
2   // Program to insert numbers into a sorted array.
3   #include <iostream>
4   using namespace std ;
5
6   bool insert( int array[], int &num_els, int array_size, int new_number ) ;
7   // Purpose      : Inserts a number into a sorted array.
8   // Precondition : array is a sorted array,
9   //                num_els is number of elements in the array,
10  //                array_size is the size of the array,
11  //                new_number is the number to insert into the array.
12  // Postcondition: array is a sorted array containing new_number,
13  //                num_els is the updated number of elements in the
14  //                array.
15  //                The return value is true if the number is
16  //                successfully inserted into the array;
```

```
17  //                    otherwise the return value is false.
18
19  int main( void )
20  {
21    const int MAX_SIZE = 20 ;
22    int sorted_array[MAX_SIZE] ;
23    int number, current_size = 0 ;
24    char more ;
25    bool success ;
26
27    // Enter numbers into the array until either the array is full
28    // or there are no more numbers to enter.
29    do
30    {
31      cout << "Enter the number to insert : " ;
32      cin >> number ;
33      success = insert( sorted_array, current_size, MAX_SIZE, number ) ;
34      if ( success )
35      {
36        cout << "Array after inserting " << number << ": ";
37        for ( int i=0; i<current_size; i++ )
38          cout << sorted_array[i] << ' ' ;
39        cout << endl ;
40        do
41        {
42          cout << "Any more numbers to insert( y or n )? " ;
43          cin >> more ;
44        }
45        while ( more !='n' && more !='N' && more != 'y' && more != 'Y' ) ;
46      }
47      else
48      {
49        cout << "Failed to add "<< number << ": array is full" << endl ;
50        more = 'n' ;
51      }
52    }
53    while ( more == 'y' || more == 'Y' ) ;
54    return 0 ;
55  }
56
57  bool insert(int array[], int &num_els, int array_size, int new_number)
58  {
59    int pos = 0 ;
60    // array full?
61    if (num_els==array_size) return false ;
62    // first number to be added to the array?
63    if ( num_els == 0 )
64      array[0] = new_number ;
65    else
66    { // search for the position of the new number
```

```
67    while ( pos < num_els && new_number > array[pos] ) pos++ ;
68    // position found, shift existing numbers
69    for ( int i = num_els ; i > pos ; i-- )
70      array[i] = array[i-1] ;
71    // insert the new number into the array
72      array[pos] = new_number ;
73    }
74    num_els++ ;
75    return true ;
76 }
```

A sample run of this program is:

```
Enter the number to insert : 7
Array after inserting 7: 7
Any more numbers to insert( y or n )? y
Enter the number to insert : 2
Array after inserting 2: 2 7
Any more numbers to insert( y or n )? y
Enter the number to insert : 9
Array after inserting 9: 2 7 9
Any more numbers to insert( y or n )? y
Enter the number to insert : 4
Array after inserting 4: 2 4 7 9
Any more numbers to insert( y or n )? y
Enter the number to insert : 8
Array after inserting 8: 2 4 7 8 9
Any more numbers to insert( y or n )? n
```

7.9 Passing a multi-dimensional array to a function
（向函数传递多维数组）

When a multi-dimensional array is passed to a function, the function parameter list must contain the size of each dimension of the array, except the first. Consider the next program, P7K, which sums the elements of a two-dimensional array.

当向函数传递多维数组时，在函数的形参列表中必须声明数组的每一维（除第一维外）的大小。下面的程序计算二维数组元素之和。

Program Example P7K

```
1  // Program Example P7K
2  // Demonstration of a two-dimensional array as a function argument.
3  #include <iostream>
4  using namespace std ;
5
6  int sum_array( const int array[][2], int no_of_rows ) ;
7  // Purpose : Sums the elements of a 2-D integer array.
8  // Parameters: A 2-D array and the number of rows in the array.
9  // Returns : The sum of the array elements.
10
11 int main( void )
12 {
13   int values[5][2] = { { 31, 14 },
14                        { 51, 11 },
```

```
15                         {  7, 10 },
16                         { 13, 41 },
17                         { 16, 18 } } ;
18   int sum ;
19
20   sum = sum_array( values, 5 ) ;
21   cout << "The sum of the array elements is " << sum << endl ;
22   return 0 ;
23 }
24
25 int sum_array( const int array[][2], int no_of_rows )
26 {
27   int total = 0 ;
28
29   for ( int row = 0 ; row < no_of_rows ; row++ )
30   {
31     for ( int col = 0 ; col < 2 ; col++ )
32       total += array[row][col] ;
33   }
34   return total ;
35 }
```

Running this program will display the following on the screen:

```
The sum of the array elements is 212
```

The elements of a two-dimensional array are stored row by row in contiguous memory locations. A sketch of memory would look like this:

二维数组元素按行存储在一块连续的存储单元中。

31	14	51	11	7	10	13	41	16	18

Any element `array[i][j]` in program P7K has an offset $i*2 + j$ from the starting memory location of `array[0][0]`. For example, `array[3][1]` (= 41) has an offset of 7 (= 3 * 2 + 1). In order to calculate the offset, the number of columns (= 2) is required; hence the need for the compiler to know the value of the second dimension on line 25.

程序P7K中的数组元素array[i][j]相对于数组起始地址的偏移量为 $i*2 + j$。例如，数组array[3][1]（=41）的偏移量为7（=3*2+1）。为了计算这个偏移量，必须知道数组的列数（=2），因此编译器要求在第 25 行必须指明数组的第二维的长度。

7.10　Passing a structure variable to a function
（向函数传递结构体变量）

When you pass a structure variable to a function, you pass a copy of the member values to that function. This means that the values in the structure variable cannot be changed within the function. As with any variable, the values in a structure variable can only be changed from within a function if the variable is passed by reference to the function. The next program demonstrates passing a structure variable by value and passing a structure variable by reference to two different functions. In the function `get_student_data()` the argument is passed by

向函数传递结构体变量时，实际传递给函数的是该结构体变量成员值的副本。这就意味着结构体变量的成员值不能在函数中修改。就像其他变量一样，仅当变量是按引用传递给函数时，结构体变量的成员值才可以在函数中修改。

reference and in `display_student_data()` the argument is
passed by value.

Program Example P7L

```cpp
1  // Program Example P7L
2  // Demonstration of using a structure as a function argument.
3  // Program reads data for a student and then displays it.
4  #include <iostream>
5  #include <iomanip>
6  using namespace std ;
7
8  void display_student_data( struct student_rec student_data ) ;
9  // Purpose : This function displays student data.
10 // Parameter: A student record structure variable.
11
12 void get_student_data( struct student_rec& student_ref ) ;
13 // Purpose : This function reads student data from the keyboard.
14 // Parameter: A reference to a student record structure variable.
15
16 struct student_rec // Student structure template.
17 {
18   int number ;
19   float scores[5] ;
20 } ;
21
22 int main( void )
23 {
24   struct student_rec student ;
25
26   get_student_data( student ) ;
27   display_student_data( student ) ;
28   return 0 ;
29 }
30
31 void display_student_data( struct student_rec student_data )
32 {
33   cout << endl << "The data in the student structure is:" << endl ;
34   cout << " Number is " << student_data.number << endl ;
35   cout << " Scores are:" ;
36   cout << fixed << setprecision( 1 ) ;
37   for ( int i = 0 ; i < 5 ; i++ )
38     cout << setw( 5 ) << student_data.scores[i] ;
39   cout << endl ;
40 }
41
42 void get_student_data( struct student_rec& student_ref )
43 {
44   cout << "Number: " ;
45   cin >> student_ref.number ;
46   cout << "Five test scores: " ;
47   for ( int i = 0 ; i < 5 ; i++ )
```

```
48      cin >> student_ref.scores[i] ;
49 }
```

A sample run of this program is:

```
Number: 1234
Five test scores: 75  80  65  45  68

The data in the student structure is:
  Number is 12345
  Scores are: 75.0 80.0 65.0 45.0 68.0
```

The structure template is declared outside `main()` on lines 16 to 20. When a structure template is defined outside `main()`, it makes the structure template *global*. This means that the structure template is known in `main()` and in the functions `display_student_data()` and `get_student_data()`. There is no need, therefore, to declare the structure template in each function.

Line 26 passes the structure variable `student` by reference to `get_student_data()`.

Line 27 passes the value of the structure variable `student` to `display_student_data()`. When a structure variable is passed by value to a function, the entire structure data must be copied to the function parameter. For a large structure, this is a significant overhead and it is preferable, therefore, to use a reference. If the argument isn't changed by the function, the `const` keyword should be used.

To avoid the overhead involved in copying the student data in program P7L, lines 8 and 31 are modified to:

结构体模板是在main( )函数之外的第16行至第20行声明的。在main( )函数之外定义结构体模板时，结构体模板是全局的。这样，该结构体模板就在所有的函数，包括main( )、display_student_data( )、get_student_data( )中都是可用的。因此，就不必在每个函数中都声明一次结构体模板了。

当一个结构体变量被按值传递给函数时，整个结构体的成员数据必须复制给函数的形参。对于一个较大的结构体，这种数据复制带来的时间开销是很大的，因此，更好的方法是采用按引用传递方式。如果不希望实参在函数中被修改，就使用const 关键字。

```
8  void display_student_data( const struct student_rec& student_data ) ;
31 void display_student_data( const struct student_rec& student_data)
```

7.11　Passing a string to function
（向函数传递字符串）

The same considerations should be kept in mind when using strings as arguments as when using structure variables as arguments, i.e. passing a string by value means copying all the characters of the string to a function parameter. To avoid this overhead it is preferable to pass strings by reference.

切记使用字符串作为函数实参和使用结构体变量作为函数实参需要考虑的问题是一样的。也就是说，按值传递字符串意味着将字符串中的所有字符都复制给函数形参。为了避免这一时间开销，更好的方式是按引用传递字符串。

7.11.1　Passing a C++ string to a function

The next program demonstrates passing a C++ string by `const` reference to a function that counts the number of vowels in the string.

Program Example P7M

```
1    // Program Example P7M
2    // Demonstration of using a C++ string as a function argument.
3    #include <iostream>
4    #include <string>
5    using namespace std ;
6
7    int vowel_count( const string& str ) ;
8    // Purpose : Finds the number of vowels in a C++ string.
9    // Parameter: A C++ string.
10   // Returns : The number of vowels in the string.
11
12   int main( void )
13   {
14     string s = "This string contains vowels" ;
15     int n = vowel_count( s ) ;
16     cout << "The number of vowels in \"" << s <<"\" is "<< n << endl ;
17     return 0 ;
18   }
19
20   int vowel_count( const string& str )
21   {
22     int str_len = str.length() ;
23     char ch ;
24     int vowel_count = 0 ;
25
26     for ( int i = 0 ; i< str_len ; i++ )
27     {
28       ch = str.at( i ) ;
29       if ( ch == 'A' || ch == 'a' ||
30            ch == 'E' || ch == 'e' ||
31            ch == 'I' || ch == 'i' ||
32            ch == 'O' || ch == 'o' ||
33            ch == 'U' || ch == 'u' )
34         vowel_count++ ;
35     }
36     return vowel_count ;
37   }
```

The output from this program is:

```
The number of vowels in "This string contains vowels" is 7
```

7.11.2 Passing a C-string to a function

To use C-strings instead of C++ strings in program P7M, the
following modifications must be made to the program:

Since a C-string is an array of characters, line 14 changes to:

```
  char s[] = "This string contains vowels" ;
```

The function argument s on line 15 is now a character array. This
means that line 19 must change to:

```
int vowel_count( const char str[] )
```

The prototype on line 7 must also change to correspond to the function header on line 20.

```
int vowel_count( const char str[] ) ;
```

Within the function `vowel_count()`, two changes are required. Line 22 changes to:

```
int str_len = strlen( str ) ;
```

and line 28 changes to:

```
ch = str[ i ] ;
```

7.12 Recursion（递归）

Recursion is a programming technique in which a problem can be defined in terms of itself. The technique involves solving a problem by reducing the problem to smaller versions of itself. For example, the factorial of a positive integer is the product of the integers from 1 through to that number. For example, 3 factorial (written as 3!) is 3*2*1 which is also 3*2!

The mathematical definition of factorial of a number n is:

$$n! \text{ is } \begin{cases} 1 & \text{when } n \text{ is } 0 \\ n * (n-1)! & \text{when } n > 0 \end{cases}$$

From this definition,

(a) 0!=1. This is called the *base case*.

(b) For a positive integer *n*, factorial *n* is *n* times the factorial of *n*−1. This is called the *general case* and clearly indicates that factorial is defined in terms of itself and is therefore an example of recursion.

递归是一种根据问题本身来定义问题的编程技术，它通过将问题分解为与其自身相同的、只是规模小一些的子问题来解决问题。例如，一个正整数的阶乘就是从1一直到该正整数的所有整数的乘积，例如，3的阶乘（记为3!）是3*2*1，也可以写成3*2!。

(a) 0! = 1，这被称为基线情况。

(b) 正整数n的阶乘是n乘以n−1的阶乘。这被称为一般情况，显然阶乘是根据其自身来定义的问题。因此，它是一个典型的递归的例子。

Using the definition, factorial 3 is calculated as follows:

- The value of n is 3 so, using (b) above, 3! = 3 * 2!
  - Next find 2! Here n = 2 so, using (b) again, 2! = 2 * 1!
    - Next find 1! Here n = 1 so, using (b) again, 1! = 1 * 0!
      - Next find 0! In this case using (a), 0! is defined as 1.
    - Substituting for 0! gives 1! = 1 * 1 = 1.
  - Substituting for 1! gives 2! = 2 * 1! = 2 * 1 = 2.
- Finally, substituting for 2! gives 3! = 3 * 2! = 3 * 2 = 6.

The next program calculates the factorial of any positive integer.

Program Example P7N

```
1   // Program Example P7N
2   // Program to compute the factorial of a number using recursion.
3   #include <iostream>
4   using namespace std;
5
```

```
6  int main( void )
7  {
8    unsigned int factorial( int n ) ;
9    unsigned int fact_n ;
10   int n ;
11
12   do // Read a number from the keyboard
13   {
14     cout << "Enter zero or a positive number " ;
15     cin >> n ;
16   }
17   while ( n < 0 ) ;
18
19   fact_n = factorial( n ) ;
20   cout << "Factorial " << n << " is " << fact_n << endl ;
21   return 0 ;
22 }
23
24 unsigned int factorial( int n )
25 // Purpose : Recursive function to calculate n!
26 // Parameter: The number for which the factorial is required.
27 // Returns : n!
28 {
29   if ( n == 0 )
30     return 1 ; // Base case
31   else
32     return ( n * factorial(n-1) ) ; // Function calls itself
33 }
```

Note that

- Every recursive function must have at least one base case which stops the recursion and
- A general case that eventually reduces to a base case.

The factorial function could also be written using iteration, i.e. using a loop.

注意：

- 每一个递归函数必须至少有一个基线情况，它是用来结束递归过程的。
- 一般情况必须最终能转化为基线情况。

```
unsigned int factorial( int n )
// Purpose  : Iterative function to calculate n!
// Parameter: The number for which the factorial is required.
// Returns  : n!
{
  unsigned int  fact ;
  int i ;

  fact = 1 ;
  for ( i = 2 ; i <= n ; i++ )
    fact *= i ;

  return fact ;
}
```

The recursive version will execute more slowly than the iterative equivalent because of the added overhead of the function calls in line 32. The advantage of the recursive version, on the other hand, is that it is clearer because it follows the actual mathematical definition of factorial.

递归方法编写的程序的执行效率低于迭代方法编写的程序的执行效率，这是因为增加了程序第32行函数调用的开销。不过，由于递归程序遵循了阶乘的实际数学定义，因此用递归方法编写的程序具有更清晰的优点。

7.13　Function overloading（函数重载）

Function overloading is used when there is a need for two or more functions to perform similar tasks, but where each function requires a different number of arguments and/or different argument data types. Program P7I used the function sum_array() to calculate the sum of the elements in a one-dimensional array and program P7K used a different function with the same name, sum_array(), to calculate the sum of the elements in a two-dimensional array.

函数重载通常用在两个或多个函数需要执行相同功能，而每个函数需要不同数量的实参或者实参数据类型不同的情况下。

If required, both versions of sum_array() can be used in a program. Using different functions with the same name in a program is called *function overloading* and the functions are called *overloaded functions*. Function overloading requires that each overloaded function have a different parameter list, i.e. a different number of parameters or at least one parameter with a different data type.

在一个程序中，使用具有相同函数名的不同函数称为函数重载，这样的函数称为重载函数。函数重载要求每个重载函数具有不同的形参列表，即形参个数不同，或者至少有一个形参类型不同。

Program Example P7O

```
1   // Program Example P7O
2   // Demonstration of function overloading.
3   #include <iostream>
4   using namespace std ;
5
6   int sum_array( const int array[], int no_of_elements ) ;
7   // Purpose   : Sums the elements of a 1-D integer array.
8   // Parameters: An array and the number of elements in the array.
9   // Returns   : The sum of the array elements.
10  int sum_array( const int array[][2], int no_of_rows ) ;
11  // Purpose   : Sums the elements of a 2-D integer array.
12  // Parameters: A 2-D array and the number of rows in the array.
13  // Returns   : The sum of the array elements.
14
15  int main( void )
16  {
17    int one_d_array[5] = { 0, 1, 2, 3, 4 } ;
18    int sum ;
19
20    sum = sum_array( one_d_array, 5 ) ;
21    cout << "The sum of the 1-D array elements is "
22         << sum << endl ;
23
24    int two_d_array[3][2] = { { 0, 1 },
25                              { 11, 12 },
26                              { 21, 22 } } ;
```

```
27
28   sum = sum_array( two_d_array, 3 ) ;
29   cout << "The sum of the 2-D array elements is " << sum << endl ;
30   return 0 ;
31 }
32
33 int sum_array( const int array[], int no_of_elements )
34 {
35   int total = 0 ;
36
37   for ( int index = 0 ; index < no_of_elements ; index++ )
38     total += array[index] ;
39   return total ;
40 }
41
42 int sum_array( const int array[][2], int no_of_rows )
43 {
44   int total = 0 ;
45
46   for ( int row = 0 ; row < no_of_rows ; row++ )
47   {
48     for ( int col = 0 ; col < 2 ; col++ )
49         total += array[row][col] ;
50   }
51   return total ;
52 }
```

The output from this program is:

```
The sum of the 1-D array elements is 10
The sum of the 2-D array elements is 67
```

The compiler decides which of the two sum_array() functions to call based on matching arguments with parameters. This means that line 20 calls sum_array() on lines 33 to 40, because the number of arguments and their data types on line 20 match the number of parameters and their data types on line 33, i.e. a one-dimensional array and an integer. Similarly, line 28 calls sum_array() on lines 42 to 52, because the argument types on line 28 and parameter types on line 42 are a two-dimensional array and an integer.

编译器根据实参与形参匹配的结果来确定调用两个sum_array( )函数中的哪一个。

7.14 Storage classes `auto` and `static`
（auto 和static 存储类型）

7.14.1 `auto`

The variables defined inside a function are auto (automatic) by default. Every time a function (including main()) is entered, storage for each auto variable is allocated. When the function is completed, the allocated storage is freed, and any values in the auto variables

在函数内部定义的变量是auto（自动）存储类型。每次进入函数［包括main( )］时，都为每个auto存储类型的变量重新分配内存空

are lost. Such variables are known as *local* variables and are known only within the function in which they are defined. If you do not specify a storage class, `auto` is assumed by default.

For example:

```
void any_function()
{
  auto int var1 ;
  auto float var2[10] ;
  ...// Body of function follows.
}
```

The variables `var1` and `var2` have been defined with the storage class `auto`. As `auto` is the default, the keyword `auto` may be omitted. Automatic variables permit efficient use of storage, because the storage used by `auto` variables is released for other purposes when a function terminates.

间；函数执行结束时，将分配的内存空间释放，存储在auto变量中的任何数值都将丢失。这样的变量称为局部变量，只能在定义它的函数内部访问。如果没有指定变量的存储类型，那么变量的存储类型就默认为auto。

当函数执行结束后，自动变量占用的存储空间被释放用于其他用途，因此自动变量有利于提高存储空间的利用率。

7.14.2 `static`

`static` variables, like `auto` variables, are local to the function in which they are defined. However, unlike `auto` variables, `static` variables are allocated storage only once and so retain their values even after the function terminates.

The next program illustrates the differences between `auto` and `static` variables.

和auto变量一样，static变量对定义它们的函数而言，也是局部的。但不同于auto变量的是，只为static变量分配一次存储空间，因此即使函数执行结束后，仍然保持其值不变。

Program Example P7P

```
1   // Program Example P7P
2   // Demonstration of the difference between static and auto.
3   #include <iostream>
4   using namespace std ;
5
6   void any_func( void ) ;
7
8   int main( void )
9   {
10    for ( int i = 0 ; i < 10 ; i++ ) // Call a function ten times.
11      any_func() ;
12    return 0 ;
13  }
14
15  void any_func( void )
16  {
17    int auto_var = 0 ;
18    static int static_var = 0 ;
19
20    static_var++ ;  // Increment the static variable.
21    auto_var ++ ;   // Increment the auto variable.
22    cout << "auto_var = " << auto_var << " "
```

```
23              << "static_var = " << static_var << endl ;
24 }
```

The output from this program is:

```
auto_var = 1 static_var = 1
auto_var = 1 static_var = 2
auto_var = 1 static_var = 3
auto_var = 1 static_var = 4
auto_var = 1 static_var = 5
auto_var = 1 static_var = 6
auto_var = 1 static_var = 7
auto_var = 1 static_var = 8
auto_var = 1 static_var = 9
auto_var = 1 static_var = 10
```

The output from this program shows that the variable static_var was initialised only once and retained its value between each function call, i.e. the value of static_var was incremented every time any_func() was called. In contrast, the auto variable auto_var was created and initialised to 0 every time the function was entered and then incremented to 1.

程序的输出结果表明，变量 static_var只进行了一次初始化，并且在每次函数调用时都能保留其原有的值，即每次调用函数 any_func( )之后，变量static_var 的值都会被加1。相反，自动变量 auto_var在每次进入函数时都被重新创建并初始化为0，然后增加到1。

7.15   The scope of a variable（变量的作用域）

The scope of a variable refers to the part of the program in which a variable can be accessed. There are two types of scope: *block scope* and *global scope*.

变量的作用域指的是程序中可以访问到变量的部分，有两种作用域类型：块作用域和全局作用域。

7.15.1   Block scope

A *block* is one or more statements enclosed in braces { and } that also includes variable declarations. A variable declared in a block is accessible only within that block.

The following program segment illustrates block scope:

一个语句块就是位于花括号{和} 之内包含变量声明的一条和多条语句。在一个语句块中声明的变量只能在该语句块中被访问。

```
void f( int x ) ;

int main ( void )
{
  float f = 0 ;
  ...
  if ( f > 0 )
  {
    // f is accessible everywhere in this block.
    char c ; // c is accessible from here to the end of this block.
    if ( f == 1 )
    {
      double d ; // d is accessible here
      ...
    } // d is destroyed.
```

```
    // f and c are accessible, d is not.
    ...
    } // c is destroyed at the end of the block.
    // f is still accessible here, but c is not.
  ...
  return 0 ;
} // f is destroyed at the end of the program.

void f( int x )
{
  // x is accessible here.
  int y ;
  ...
  if ( x == 1 )
  {
    int z ;
    // x, y and z are accessible here.
    ...
  } // z is destroyed here.
  // x and y are accessible here, but z is not.
  ...
} // x and y are destroyed when the function terminates.
```

Variables declared inside the parentheses of a `for` are accessible within the parentheses, as well as in the statement(s) contained in the `for` loop.

在for的圆括号内声明的变量可以在圆括号内被访问，还可以在for循环体语句内被访问。

For example:

```
for ( int i = 0 ; i < 10 ; i++ ) // i is declared inside the ().
{                                 // i is accessible inside the ().
  // i is also accessible here.
  ...
  cout << i ;
  ...
} // i is destroyed at the end of the block.
// i is no longer accessible.
...
for ( int j = 0 ; i < 10 ; i++ )
  cout << j ; // The for loop controls only 1 statement.
// j is destroyed and is no longer accessible.
...
```

7.15.2 Global scope

A variable declared outside `main()` is accessible from anywhere within the program and is known as a global variable. For example:

在main( )之外声明的变量称为全局变量，它可以在程序的任何位置被访问。

```
int g ; // g is a global variable.

void f1( void ) ;
void f2( void ) ;
```

```
int main( void )
{
  int a ;
  // a and g are accessible here.
  ...
return 0 ;
} // Program ends, a and g are destroyed.

void f1( void )
{
  int b ;
  // b and g are accessible here.
  ...
} // Function ends, b is destroyed.

void f1( void )
{
  // g is accessible here.
  ...
}
```

Because global variables are known, and therefore can be modified, within every function, they can make a program difficult to debug and maintain. Global variables are not a substitute for function arguments. Strictly speaking, apart from its own local variables, a function should have access only to the data specified in the function parameter list.

由于全局变量可以在任何函数中被访问，因此也可以被任何函数所修改，这给程序的调试和维护带来困难。全局变量不是函数实参的替代品。严格地讲，除了局部变量，函数应该只能访问函数形参列表中指定的数据。

7.15.3   Reusing a variable name

It is permissible to give a variable the same name as another variable in another block. This is known as *name reuse*.

The next program demonstrates name reuse and shows its effect on the scope of a variable.

C语言允许不同语句块中的两个变量具有相同的名字，这称为变量名重用。

Program Example P7Q

```
1  // Program Example P7Q
2  // Demonstration of variable scope and name reuse.
3  #include <iostream>
4  using namespace std ;
5
6  int i = 1 ; // i is a global variable.
7
8  void f( void ) ;
9
10 int main( void )
11 { // Start of program block.
12
13   cout << "Global variable i = " << i << endl ;
14
15   int i = 2 ; // i is reused here.
```

```
16      cout << "Variable i declared in main() = " << i << endl ;
17
18      // The global variable i can be accessed by using ::
19      cout << "Global variable i = " << ::i << endl ;
20
21      char c = 'x' ;
22      cout << "Variable c declared in main() = " << c << endl ;
23      { // Start of a statement block.
24        int c = 3 ; // c is reused.
25
26        cout << "Variable c declared in statement block = " << c << endl ;
27        // variable c declared in main() is not accessible here.
28      } // End of statement block.
29
30      cout << "Variable c declared in main() = " << c << endl ;
31      // Variable c declared in the statement block is not accessible here.
32
33      f() ;
34
35      cout << "Variable i declared in main() = " << i << endl ;
36
37      cout << "Global variable i = " << ::i << endl ;
38
39      return 0 ;
40
41 } // End of program block.
42
43 void f( void )
44 { // Start of a function block.
45
46      cout << "Global variable i = " << i << endl ;
47
48      char i = 'y' ;
49      cout << "Variable i declared in f() = " << i << endl ;
50
51      cout << "Global variable i = " << ::i << endl ;
52
53 } // End of function block.
```

The output from this program is:

```
Global variable i = 1
Variable i declared in main() = 2
Global variable i = 1
Variable c declared in main() = x
Variable c declared in statement block = 3
Variable c declared in main() = x
Global variable i = 1
Variable i declared in f() = y
Global variable i = 1
Variable i declared in main() = 2
Global variable i = 1
```

If a variable is declared in an inner block and if a variable with the same name is declared in a surrounding block, the variable in the inner block *hides* the variable of the surrounding block. For example, the declaration of `i` on line 15 hides the global variable `i` declared on line 6. Similarly, the declaration of `c` on line 24 hides the declaration of `c` on line 21.

If a global variable is hidden by a local variable, the global variable can still be accessed using the unary scope resolution operator :: (two colons), as shown on line 19.

如果在一个内层的语句块内声明了一个变量，在包含这个语句块的外层语句块内又声明了另一个与其同名的变量，那么内层语句块内的变量将屏蔽外层语句块内的变量。

如果一个全局变量被局部变量所屏蔽，那么可以通过一元作用域运算符::（两个冒号）来访问这个全局变量，如第19行所示。

7.16   Mathematical functions（数学函数）

To use any of the mathematical functions place the statement

```
#include <cmath>
```

at the start of the program.

7.16.1   Some trigonometric functions

Table 7.1 shows some trigonometric functions.

Table 7.1   Some trigonometric functions

Function	Description
cos(x)	Cosine of angle x in radians. x is a double value.
	Returns a double value.
sin(x)	Sine of angle x in radians. x is a double value.
	Returns a double value.
tan(x)	Tangent of angle x in radians. x is a double value.
	Returns a double value.

The next program demonstrates `sin()`, `cos()` and `tan()` functions.

Program Example P7R

```
1   // Program Example P7R
2   // Demonstration of the functions sin(), cos(), and tan().
3   #include <iostream>
4   #include <iomanip>
5   #include <cmath>
6   using namespace std ;
7
8   int main( void )
9   {
10     const double RADIANS_IN_A_DEGREE = 57.29578 ;
11
12     double degrees, radians ;
13
14     cout << "Input the angle in degrees: " ;
15     cin >> degrees ;
16     radians = degrees / RADIANS_IN_A_DEGREE ;
17     cout << fixed << setprecision( 3 )
18         << "sin(" << degrees << ")=" << sin( radians ) << endl
19         << "cos(" << degrees << ")=" << cos( radians ) << endl
```

```
20          << "tan(" << degrees << ")=" << tan( radians ) << endl ;
21    return 0 ;
22 }
```

A sample run of this program is:

```
Input the angle in degrees: 60
sin(60.000)= 0.866
cos(60.000)= 0.500
tan(60.000)= 1.732
```

7.16.2　Pseudo-random number functions

To use the pseudo-random generating functions rand() and srand(), place the statement

```
#include <cstdlib>
```

at the start of the program.

Table 7.2 shows pseudo-random number functions.

Table 7.2　Pseudo-random number functions

Function	Description
rand()	Returns a pseudo-random integer value. Each call to rand() will produce a pseudo-random integer value. However, each time the program is executed the same sequence of integer values will be returned, unless a different seed value is used with the srand() function.
srand(n)	Use this function to set the seed (starting value) for pseudorandom numbers generated by rand(). The seed value, n, is an unsigned int.

该函数返回一个伪随机整数值，每次调用rand( )都产生一个伪随机整数值，但是每次程序执行都产生相同的伪随机数序列，除非使用函数srand( )来设置不同的随机数种子。

使用该函数来为rand( )产生的伪随机数设置种子（起始值）。种子值n是一个无符号整数。

The next program demonstrates the use of the random number functions.

Program Example P7S

```
1  // Program Example P7S
2  // Demonstration of the random number functions.
3  #include <iostream>
4  #include <cstdlib>
5  #include <ctime>
6  using namespace std ;
7
8  int main( void )
9  {
10   time_t t ;  // Define t as variable of type time_t.
11
12   t = time( 0 ) ; // Current time in seconds.
13   // Use the time to initialise the random number generator.
14   srand( t ) ; // Set the seed to the time.
15   // Generate five random numbers between 0 and 20.
16   cout << "Five random numbers in the range 0-20" << endl ;
17   for ( int i = 0 ; i < 5 ; i++ )
18   {
19     int r = rand() % 21 ; // %21 ensures a number between 0 and 20.
```

```
20     cout << r << endl ;
21   }
22   return 0 ;
23 }
```

A sample run of this program is:

```
Five random numbers in the range 0-20:
9
8
17
18
17
```

The statement

```
t = time( 0 ) ;
```

on line 12 assigns to t the current time (measured in seconds since midnight on 1 January 1970, GMT) which is used as the random number seed on line 14.

Without line 14, the program displays the same sequence of random numbers every time the program is run.

用当前时间（测得的自1970年1月1日午夜开始的以秒为单位的格林尼治时间）为t赋值，它被作为第14行的随机数种子。

如果没有第14行，那么每次运行程序时，将显示相同的随机数序列。

Programming pitfalls

1. Parameters and arguments must agree in number and type. For example, if you call a function with two `int`s and a `float` as arguments, then the function must have two `int`s and a `float` as parameters.
2. Be aware of the difference between passing arguments by value and passing arguments by reference. When you pass an argument by value, the value of that argument cannot be changed from within the function. To change the value of an argument from within a function, the argument must be passed by reference to the function.
3. Parentheses `()` are used to enclose function arguments and parameters; brackets `[]` are used for the subscripts of an array.
4. There is no semicolon after a function header, but there is one after a function prototype.
5. The angle in the trigonometric functions is measured in radians, not degrees.
6. If you omit a function prototype, C++ assumes the function will return an integer value. If the function is returning a data type other than integer, the compiler will give an error message.
   Avoid the error message by always including a prototype for every function in the program.

1. 形参和实参的数目和类型必须一致。例如，如果用两个整数和一个浮点数作为实参调用一个函数，那么这个被调函数的形参也必须是两个整数和一个浮点数。
2. 注意按值传递和按引用传递之间的区别。按值传递实参时，该实参是不能在函数内改变的。要想在函数内改变实参的值，必须按引用给函数传递实参。
3. 圆括号( )通常用于将函数的实参和形参括起来，而方括号[ ]则用于将数组的下标括起来。
4. 函数头的后面没有分号，但函数原型的后面有一个分号。
5. 在三角函数中，角的单位是弧度，不是角度。
6. 如果省略函数原型，那么C++假定函数返回值为整数。如果函数返回值不是整数，那么编译器将给出错误信息提示。如果总是在程序中包含每个函数的原型，那么就可以避免出现该错误信息提示。

Quick syntax reference

	Syntax	Examples
Function prototype	```type function_name (type parameter1, type parameter2, ... type parametern) ;```	```float average (float a[], int n) ;```
Function definition	```type function_name (type parameter1, type parameter2, ... type parametern) { local variables ; executable statements ; return expression ; }```	```float average (float a[]; int n) { int i ; float sum = 0, average ; for (i = 0 ; i < n ; i++) sum += a[i] ; average = sum / n ; return average ; }```

(cont.)

	Syntax	Examples
Function call	`variable = function_name( argument1,` `argument2,` `...` `argumentn ) ;`	`float array[10], n ;` `float avg ;` `...` `avg = average( array, n ) ;`

Exercises

1. (a) Write function prototypes for the following functions:

Function	Parameter(s)	Return value
f1	int	char
f2	char	char
f3	int	int
f4	two integers	none
f5	string reference	int
f6	none	bool
f7	an integer array	float
f8	a double array	none
f9	none	none

For example, the function prototype for the function f1 () is:

```
char f1( int variable_name ) ;      or      char f1( int ) ;
```

(b) Write a statement to call each of the above functions.

For example, to call function f1 () :

```
char c_val ;
int i_val ;
c_val = f1( i_val ) ;
```

2. What is wrong with each of the following functions?

(a)
```
void max(a, b) ;
   if ( a > b)
      return a ;
   else
      return b ;
```

(b)
```
bool test(int)
{
   for(int i=1;i< n;i++)
      cout << "x";
}
```

(c)
```
float min( void )
   int a, b ;
   if ( a < b )
      return a
   return b ;
}
```

3. Find the errors in this program:

```
#include <iostream>
using namespace std ;

int f1() ;
void f2( int a ) ;

int main( void )
{
   int v1 ;
   float v2 ;
   char v3 ;
```

```
   v1 = f1[1] ;
   v2 = f1() ;
   v3 = f2() ;
   v2 = f2( 1, 2, 3 ) ;
   return 0 ;
}
void f1( int a )
{
   return a / 2 ;
}

void f2( int a, b )
{
   return a + b ;
}
```

4. What is the output from the following?

```
#include <iostream>
using namespace std ;

int f( int val1, int val2 = 0 ) ;

int main( void )
{
   int var ;

   var = f( 1, 2 ) + 1 ;
   var = f( var + 1 ) ;
   var = f( f( 1, 2 ), f( 3, var ) ) ;
   cout << "The value of var is " << var << endl ;
   return 0 ;
}

int f( int val1, int val2 )
{
   if ( val1 > val2 )
     return ( val1 - val2 ) ;
   else
     return ( val2 - val1 ) ;
}
```

5. What is the output from the following?

```
#include <iostream>
using namespace std ;

void f( int val1, int val2 = 2 ) ;
void f( string& s ) ;
void f( char c ) ;

int main( void )
```

```
{
  string str = "this is a string" ;

  f( 1 ) ;
  f( str ) ;
  f( 'a' ) ;
  return 0 ;
}

void f( int i, int j )
{
  cout << "i = " << i << " j = " << j << endl ;
}

void f( string& s )
{
  cout << "s = " << s << endl ;
}

void f( char c )
{
  cout << "c = " << c << endl ;
}
```

6. Write a function to return the minimum of three integer values.

7. (a) Write a function to return the minimum value in an integer array of ten elements.

   (b) Modify the function to take the number of elements in the array as a parameter.

8. Write a function to test whether an integer lies within a range of values. The function prototype will be:

```
bool range_test( int val, int low = 1, int high = 10 ) ;
```

   where val is the value to be tested, low is the lower value in the range, and high is the higher value in the range. The function will return the bool value true if the value is in the specified range, otherwise it will return the bool value false.

9. Write a function to convert hours, minutes and seconds to seconds.

10. Write a function to convert seconds to hours, minutes, and seconds.

11. (a) Write a function to return the minimum value in an integer array.

    (b) Overload the function in (a) with a function to return the minimum value in a floating-point array.

12. Write a function to determine the lowest two values in an integer array. Use reference variables as parameters.

13. Modify program P7J to include a function to remove an element from the sorted array.

    Include pre and post condition comments for the additional function.

14. Write a function to count the number of words in a string.

    (Assume that every word is separated by at least one whitespace character.)

15. Write a function that has four parameters: a starting number, an ending number, an increment (default 1) and the number of numbers to be displayed per line (default 10).

16. Write a function to capitalise all the letters in a string.

17. Write a function to determine the frequency of each of the vowels in a string.

18. Write a function to reverse a string.

19. A palindrome is a word that you can spell forwards or backwards. For example, the word 'madam' is a palindrome. Write a function to check whether a word is a palindrome or not. The function should return a bool value of true or false as appropriate.

20. (a) What is the output from the following?

```cpp
#include <iostream>
using namespace std ;

void any_function() ;

int main( void )
{
  int i ;
  for ( i = 0 ; i < 10 ; i++ )
    any_function() ;
  return 0 ;
}
void any_function()
{
  static int var = 10 ;
  cout << var << endl ;
  var += 10 ;
}
```

(b) What would be the effect of replacing the function any_function() with the following:

```cpp
void any_function()
{
  static int var ;
  var = 10 ;
  cout << var << endl ;
  var += 10 ;
}
```

21. Write a function that will return the letter A the first time it is called, B the second time it is called, C the third time it is called, and so on. (Hint: use a static char)

22. What does this recursive function do?

```cpp
void recur_fun( int n )
{
  cout << n ;
  if ( n == 1 )
  return ;
  recur_fun ( n - 1 ) ;
}
```

23. Write a program to generate a set of lottery numbers, using srand() and rand().

Chapter Eight
Objects and Classes
第 8 章 对象和类

8.1 What is an object? (什么是对象？)

An object is a component of a program that knows how to perform certain actions and knows how to interact with other parts of the program. An object consists of one or more data values, which define the state or properties of the object, and functions that can be applied to the object. The functions associated with an object represent what can be done to the object and how the object behaves. An object can be anything such as a person, a computer or even an intangible such as a bank account.

Consider a computer adventure game. Among the objects in the game might be:

You, the player,

Your enemies e.g. dragons

Your weapons

Obstacles such as locked doors, rivers, etc.

The prize e.g. a fair maiden

A player will have data values to represent certain attributes, e.g. the state of their health or the weapons they possess. A player must be able to perform functions such as walk, run, attack an enemy, and rescue the fair maiden to win. Behaviour between objects is very important. For example, if you fall into the river loaded with weapons, you drown. If you fatally wound an enemy, then it is out of the game and you have one less enemy to contend with.

8.2 What is a class? (什么是类？)

A class is a general category that defines:

● The characteristics that an object of that category contains. The characteristics are called *properties* or class *data members*.

● The functions that can be applied to objects of that category. The functions are also called class *member functions* or *methods*.

That is, a class defines both the type of data and the operations that

对象是程序的一个组成部分，它知道如何执行特定的操作，知道如何去和程序的其他部分进行交互。一个对象包含一个或多个数据值及应用于这个对象的一些函数，其中这些数据值定义了这个对象的状态或者属性，和对象关联的函数表示可以对该对象做些什么及该对象是如何运作的。对象可以是任何事物，例如人、计算机，甚至是像银行账户这样的无形的东西。

类是一个广义的范畴，可以定义如下：

● 这个范畴的对象所包含的特征，称为属性或类的*数据成员*。

● 应用于这个范畴的对象的函数，称为类的*成员函数*或方法。

类同时定义了数据的类型和可作

can be applied to that data. Including both the data and functions into one unit, the class, is called *encapsulation*.

用于这些数据之上的操作。类把数据和函数包含在一起，成为一个整体。这个过程称为封装。

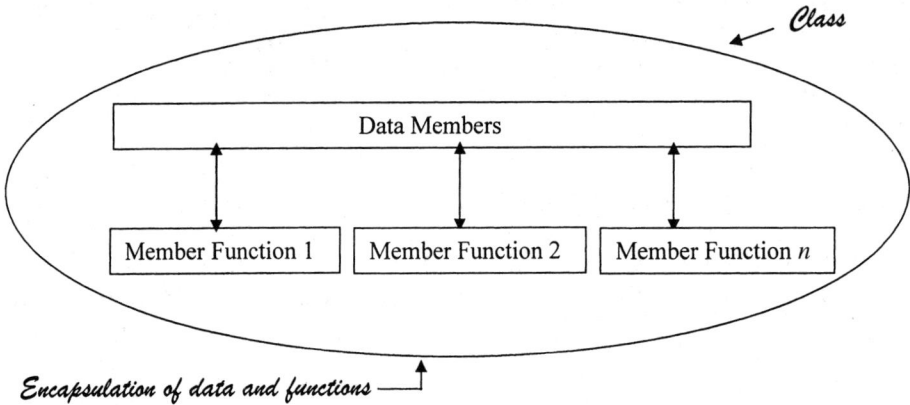

In our computer adventure game a player class may be defined as:

```
class player
{
// functions:
     walk()
     run()
     jump()
     attack()
     rescue()
     ...
//data:
     state_of_health
     type_of_weapon
     ...
} ;
```

Note: This is only a pseudo(not real)-code representation of the class used for explanatory purposes only. Details of the functions and the data types used in the class have been omitted. When writing a class in C++, these details are required.

A dragon class may be defined in pseudo-code as:

```
class dragon
{
//functions:
     walk()
     spit_fire()
     use_claws()
     use_tail()
     die()
     ...
```

```
//data:
    size
    number_of_claws
    state_of_health
    ...
};
```

There is an important distinction between a class and an instance of the class. A class is a blueprint or a template that can be used to create many instances of the class. An instance of a class is the actual object, created from the template, that can be manipulated by the member functions of the class.

The difference between a class and an instance of a class (an object) is like the difference between a noun and a proper noun. For example, person is a noun, John Smith is a proper noun that represents a specific person. Person is the class and John Smith is the object.

Getting back to the adventure game, the following statement creates an instance of a dragon:

```
class dragon george( 10, 4 ) ;
```

Here george is an instance or an object of the dragon class. The data in the parentheses represent initial values for some of the data members of george, e.g. size and number_of_claws.

If in the game george is required to spit fire, a 'message' is sent to george by calling the class member function spit_fire(). A class member function is called using the member selection operator (a dot):

```
george.spit_fire() ;
```

Any number of dragon objects can be created. For example,

```
class dragon ivan( 6, 2 ), baby( 1, 0 ) ;
```

In this example, ivan is a size 6 dragon with 2 claws and baby is a size 1 dragon with no claws.

Each dragon can perform different functions independent of each other:

```
ivan.use_claws() ;   // ivan attacks with claws.
baby.die() ;         // baby dragon dies. sorry!
george.spit_fire() ; // george attacks with fire.
```

To summarise:

- An object is an instance of a class
- An object has:
  - an *identity*, i.e. its name
  - a *state*, i.e. its data members
  - a *behaviour*, i.e. its member functions.

类和类的实例有一个非常重要的区别，类是一个蓝图或者模板，用这个模板可以创建很多个该类的实例；类的实例是从模板创建的一个实际的对象，类的成员函数可以对这个对象进行一些操作。类和类的实例（对象）之间的不同，类似于名词和专有名词之间的不同。例如，person 是一个名词，John Smith 表示一个特定的人，是一个专有名词。这里，person 就相当于类，而 John Smith 就相当于对象。

本节的主要内容可归纳为以下几点：

- 对象是类的实例。
- 一个对象具有如下特征。
  - 标识，即对象的名字；
  - 状态，即对象的数据成员；
  - 行为，即对象的成员函数。

8.3　Further examples of classes and objects
（类和对象的更进一步的示例）

8.3.1　A student class

A student is an example of a class. Some of the properties (class data members) that define a student are the name, the address, the student number, the date of birth, the course and the course mark. Some of the functions (class member functions or methods) that can be performed on a student are to update the student's course mark, display a progress report and so on. These operations are performed by the member functions of the class.

The student class can de defined in pseudo-code as:

以学生类为例，能够定义一个学生的属性（类的数据成员）包括姓名、地址、学号、出生日期、课程及课程成绩等。对一个学生可以执行的操作（类的成员函数或方法），包括更新学生的课程成绩、显示一份进展报告等。这些操作都是由类成员函数来完成的。

```
class student
{
// functions:
    update_mark()
    display_report()

// data:
    name
    address
    student_number
    date_of_birth
    course_code
    course_marks
};
```

A student object is a specific instance of the student class. Students in a course such as *John Smith* or *Liu Wei* are instances of the student class. Each of these student class objects has its own name, address, date of birth and set of marks.

8.3.2　A bank account class

In its simplest form, a bank account has an account number and a balance. These are the class data members of a bank account class. Some of the operations that can be done on a bank account class are to open the account with an amount of money and to deposit and withdraw money from the account. These operations will be represented by class member functions such as `open( amount )`, `withdraw( amount )` and `deposit( amount )`, where `amount` represents an amount of money.

在最简单的形式下，一个银行账户包含账号和余额两项信息。这两项信息都是银行账户类的数据成员。能够在银行账户类上进行的操作包括：以一定金额开户、存款和取款。这些操作分别由类成员函数 open(amount)、withdraw(amount) 和 deposit(amount) 来完成，这里 amount 代表交易的金额。

```
class bank_account
{
// functions:
    open( amount )
```

```
    deposit( amount )
    withdraw( amount )
// data:
    account_number
    balance
};
```

A bank account object is a specific instance of the bank account class. So your bank account will have a unique account number and you will have a specific amount of money in your account.

The following statement creates an instance of a bank account called `my_bank_account`:

```
class bank_account my_bank_account ;
```

To withdraw twenty pounds from `my_bank_account`:

```
my_bank_account.withdraw( 20.00 ) ;
```

8.4   Abstraction（抽象）

In the computer adventure game, each software object behaves only in some respects as its real-world counterpart. Not all the behaviour or characteristics of the real-world object is necessary in the game. For example, there may be no need for a player to eat, drink or sleep in the game. In effect, each software object is a simplification of its real world counterpart. This is called *abstraction*.

Abstraction is also used in the bank account class. The bank account class as described above doesn't describe all the characteristics and all the operations of a real bank account, but only the characteristics and operations that are relevant for the purposes of the program.

The relevant characteristics and functions of a real bank account have been abstracted to produce a software model of its real world counterpart.

Abstraction is used in our everyday lives. For example, a lecture timetable will contain a subject, a lecturer's name, a room number and a time. On the timetable, the room where the lecture is to be held is represented simply by a number. For the purpose of a student's timetable, details of the room, such as its size, are unnecessary and are therefore not given. However, for a college administrator the size of the room is relevant and so would be included in an abstraction for the purposes of college administration.

A more common example is our use and understanding of a car. It is well known that the engine makes the car go, but technical knowledge of the engine is not necessary in order to drive the car. To drive a car it is necessary to know how the various controls such

在计算机探险类游戏中，每个对象仅仅在一些方面像现实世界中的原型那样运作。现实对象的行为和特征在游戏中并非都要面面俱到。例如，游戏玩家就不必吃饭、喝水或者睡觉。事实上，游戏中的每个对象都是现实中原型的一个简化，称为抽象。

抽象同样也适用于银行账户类。前面设计的银行账户类并没有描述实际银行账户的全部特征和操作，只描述了与程序实现目的相关的那部分特征和操作。

从现实银行账户中抽取出来的、与程序相关的特征和操作就被抽象成了现实银行账户的一个软件模型。

抽象也可以应用在日常生活中。例如，一个课程表包含课程名、教师姓名、教室号、上课时间。在这个课程表里，只是简单地用一个数字来表示教室。对于一个学生的课程表来说，教室的大小等细节可以不必给出。但是对于一个大学的管理者而言，在抽象时就不能忽略教室的大小。

举一个更普遍的例子，即关于对汽车的使用和理解的。众所周知，汽车是靠发动机来驱动的，但是为了开车，人们没有必要去了解发动机的技术细节，只需知

as the gear stick, accelerator and brake are used, but what happens under the bonnet is not entirely necessary. Our understanding of a car is a simplification or abstraction of the real thing.

道怎么使用各种控制装置就可以了，例如变速换挡、油门、刹车等，至于发动机罩下发生了什么则不必去关心。人们对汽车的这种理解就是对实际事物的一种简化或抽象。

8.5　Constructing a class in C++
（用C++构造一个类）

The bank account class of section 8.3.2 can be constructed in C++ as follows:

Program Example P8A

```
1   // Program Example P8A
2   // Demonstration of a C++ class.
3   #include <iostream> // Required later for input-output.
4   #include <iomanip>  // Required later for manipulators.
5   using namespace std ;
6
7   class bank_account
8   {
9   public:
10    void open( int acc_no, double initivance ) ;
11    void deposit( double amount ) ;
12    void withdraw( double amount ) ;        The class member function prototypes.
13    void display_balance( void ) ;
14  private:
15    int account_number ;
16    double balance ;                         The class data members.
17  } ;  ← Don't forget the semicolon.
18
```

On line 7, following the keyword `class`, the class is given the name `bank_account`, which can be any valid C++ identifier.

Lines 10 to 13 declare the member functions of the class and lines 15 and 16 declare the data members of the class.

The members of the class are divided into `private` and `public` members. The keywords `private` and `public` specify the access control level for the data and function members of the class. Data members declared with `private` access control are accessible only to member functions of the class and unavailable to any functions that are not members of the class. This is called *information hiding* and prevents the data from being changed except from within the class. The `private` data members of a class can only be accessed through the `public` functions of the class.

Members declared with `public` access are accessible in any part of a program. In the `bank_account` class all the member functions are `public`, but a class member function can also have `private` access control. A `public` member function can be called from any part of a program, while a `private` member function can only be

类的成员有私有成员和公有成员两种。用关键字private和public指定类的数据成员和成员函数的访问控制权限。被声明为private的数据成员只能被这个类的成员函数访问，其他的非成员函数无权访问，这称为信息隐藏。其目的是防止数据在这个类外被修改。类的私有数据成员只能通过这个类的公有函数来访问。

声明为public的数据成员可以在程序的任何部分被访问。在bank_account类中，所有的成员函数都是公有的。当然，类成员函数也可以是私有的。公有成员函数可以在程序的任何部分被调用，而私有成员函数只能在同一个类的成员函数中被调用。公有成员函数称为类的公共接口。

called from within member functions of the same class. The public member functions are known as the *public interface* of the class.

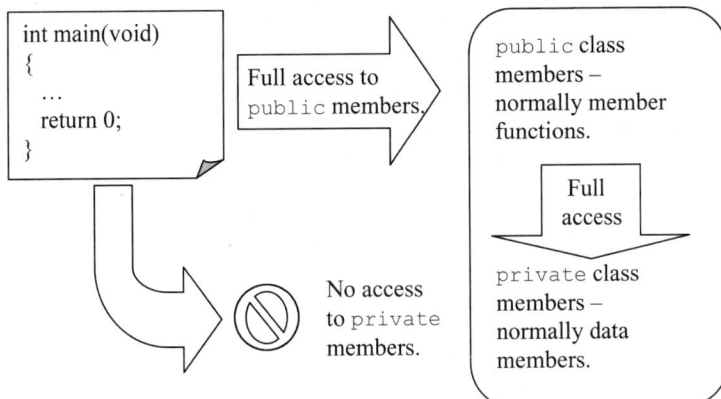

The general format of a class is:

```
class class_name
{
public:
// Details of the public interface of the class.
private:
// Private member functions and data members.
} ;
```

Generally, data members of a class are all `private` and the member functions of the class are all `public`.

Although not mandatory, the `public` section is usually placed at the start of the class before the `private` section. A programmer who wants to know how to use the class need only read the first part of the class, i.e. the public interface, to see what member functions are available and how to use them.

一般而言，类的数据成员都声明为私有成员，而成员函数都声明为公有成员。

尽管没有强制性的规定，但通常把公有成员放在类的开始部分，位于私有成员前面。程序员想要使用这个类，只需看这个类的开始部分（即公共接口），就知道哪些成员函数是可以使用的及如何使用它们。

To complete the class, the class member functions must be defined.

```
19 void bank_account::open( int acc_no, double initial_balance )
20 {
21   account_number = acc_no ;
22   balance = initial_balance ;
23 }
24
25 void bank_account::deposit( double amount )
26 {
27   balance += amount ;
28 }
29
30 void bank_account::withdraw( double amount )
31 {
32   balance -= amount ;
33 }
```

The class member function definitions.

```
34
35 void bank_account::display_balance( void )
36 {
37    cout << "Balance in Account " << account_number << " is "
38         << fixed << setprecision( 2 )
39         << balance << endl ;
40 }
41
```

A class member function has the general format:

```
return_type class_name::function_name( parameter list )
{
// member function statements.
}
```

The scope resolution operator :: is used here to specify that a function is a member of a class.

8.6 Using a class: defining and using objects
（使用类：定义和使用对象）

To use the `bank_account` class, place the class and member function definitions before `main()`. The objects of the class are defined and used in `main()`.

正确使用bank_account类的方法是，先在main( )函数之前定义这个类和它的成员函数，然后在main( )函数中定义和使用这个类的对象。

Program Example P8A...continued

```
42 int main( void )
43 {
44   class bank_account my_account ; // my_account is an object
45                                  // of class bank_account.
46
47   my_account.open( 123, 10.54 ) ; // Open account 123 with 10.54
48
49   my_account.display_balance() ;  // Display account details.
50
51   my_account.deposit( 10.50 ) ;   // Deposit 10.50
52
53   my_account.display_balance() ;
54
55   my_account.withdraw( 20.04 ) ;  // Withdraw 20.04
56
57   my_account.display_balance() ;
58   return 0 ;
59 }
```

Line 44 creates a `bank_account` object called `my_account.`:

my_account	
account_no	balance
?	?

Line 44 can also be written as:

```
bank_account my_account ;   // The keyword class is optional.
```

Although `my_account` has been created, the data members `account_no` and `balance` have not been initialised, so their values are unknown.

Line 47 initialises the data members of the class by calling the member function `open()` with values for the `account_no` and `balance`:

my_account	
account_no	balance
123	10.54

Running this program will display the following:

```
Balance in Account 123 is 10.54
Balance in Account 123 is 21.04
Balance in Account 123 is 1.00
```

8.7 Abstract data types（抽象数据类型）

C++ has a set of built-in data types such as `char`, `int` and `float`. Each one of these data types has a unique range of allowable values and a set of allowable operations and functions. For example, the `float` data type has a range of positive and negative values which are different from the range of numbers that can be held with an `int` data type (see Appendix D). An allowable operation on both the `int` and `float` data types is the square root function `sqrt()` (see program P7A). A `string` function such as `length()` (see program P6J) is not applicable to either `float` or `int` and is consequently not allowed. The details of how negative numbers are stored or how many bits of storage are used to store a value are not necessary in order to use a `float` or an `int` type variable. All that is required to know is what functions and operations can be used with a particular data type, and what is the range of allowable values for that data type. In other words, the implementation details of the data types are hidden from the programmer. This is called *data abstraction*.

This is very similar to the way in which the bank account class is used in program P8A. The bank account class defines a new data type that is not built into the C++ language. The new data type is called an *abstract data type* (ADT). Once written, to use the new bank account data type it is not necessary to know the details of how each of the class member functions work. It is only necessary to know how to call the public member functions of the class such as `open()`, `deposit()` and `withdraw()`.

C++包含一组内置的数据类型，例如字符型、整型和浮点型。每一种数据类型都有一个唯一的允许的取值范围及一组允许的操作和函数。例如，浮点型所允许的正负取值范围与整型所允许的取值范围就是不同的（见附录D）。平方根函数sqrt( )（见程序P7A）既适用于整型数据，也适用于浮点型数据。像length( )（见程序P6J）这样的字符串处理函数就不能应用于整型或浮点型数据。要使用一个浮点型或整型的变量，不需要知道负数是怎样存储的或者用多少位去存储一个数据值这些细节。对于一个特定的数据类型，只需了解哪些函数和操作可以使用及这个数据类型的合法取值范围是什么。换句话说，数据类型的实现细节对程序员是隐藏的，这称为数据抽象。

这和程序P8A中银行账户类的使用方式很类似。银行账户类定义了一种新的非内置的数据类型，这种新的数据类型称为抽象数据类型（ADT）。一旦编写完毕，要使用这个新的银行账户数据类型，则不需要了解这个类的每个成员函数是如何工作的，只要

In effect, a class object can be regarded as a "black box" where data is entered and results produced without knowing what's happening inside the box.

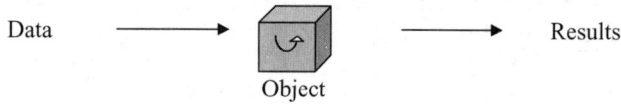

In effect, a class object can be regarded as a "black box" where data is entered and results produced without knowing what's happening inside the box.

Data ⟶ Object ⟶ Results

知道怎样去调用这个类的公有成员函数［如 open( )、deposit( ) 和 withdraw( ) 等］即可。

实际上，可以将一个类的对象看成一个"黑箱子"，把数据输入这个箱子中，然后从箱子中取结果，而不必知道箱子内部是如何工作的。

8.8　Constructors（构造函数）

Constructors play a critical role in the automatic initialisation of data. A constructor is a class member function that has the same name as the class, and is automatically called when an object of the class is created. A constructor is frequently used to provide the private data members of the object with initial values.

In the bank account class, which doesn't include a constructor, the private data members `account_number` and `balance` are assigned their initial values in the member function `open()`. If `open()` is not explicitly called from `main()` then the private data members do not have initial values and the rest of the program will give spurious results. By including a constructor in a class, it is possible to ensure that the data members of the class are automatically initialised.

In the next program, a constructor is added to the `bank_account` class in place of the `open()` member function.

构造函数在数据的自动初始化中发挥着重要的作用。构造函数是一个函数名和类名相同的类成员函数。当这个类的对象被创建时，构造函数会被自动调用。构造函数经常用来实现对象的私有数据成员初始化。

在银行账户类中，没有包含构造函数，私有数据成员 account_number 和 balance 的初值是通过成员函数 open( ) 来赋值的。如果在 main( ) 函数里没有显式地调用函数 open( )，那么私有数据成员就没有初值，其后的程序就会输出一个不合逻辑的结果。通过在类中包含构造函数，就可以保证这个类的数据成员被自动初始化。

Program Example P8B

```
1   // Program Example P8B
2   // Demonstration of a class constructor.
3   #include <iostream>
4   #include <iomanip>
5   using namespace std ;
6
7   class bank_account
8   {
9   public:
10     bank_account( int acc_no, double initial_balance ) ;
11     void deposit( double amount ) ;
12     void withdraw( double amount ) ;
13     void display_balance( void ) ;
14   private:
15     int account_number ;
16     double balance ;
17   } ;
18
19   bank_account::bank_account( int acc_no, double initial_balance )
20   {
21     account_number = acc_no ;
22     balance = initial_balance ;
```

The class constructor has the same name as the class and has no return type.

```
23 }
24
25 void bank_account::deposit( double amount )
26 {
27   balance += amount ;
28 }
29
30 void bank_account::withdraw( double amount )
31 {
32   balance -= amount ;
33 }
34
35 void bank_account::display_balance( void )
36 {
37   cout << "Balance in Account " << account_number << " is "
38        << fixed << setprecision( 2 )
39        << balance << endl ;
40 }
41
```

The class constructor on line 19 has the same name as the class and has no return type.

The constructor is never explicitly called and therefore cannot return a value. That's why a class constructor has no return type, not even `void`. Rewriting `main()` to make use of the class constructor:

这个构造函数永远不会被显式地调用，所以不能返回一个数据值。也正因为如此，类构造函数没有返回类型，甚至连void类型都没有。

```
42 int main( void )
43 {
44   bank_account my_account( 123, 10.54 ) ;
45
46   my_account.display_balance() ;
47
48   my_account.deposit( 10.50 ) ;
49
50   my_account.display_balance() ;
51
52   my_account.withdraw( 20.04 ) ;
53
54   my_account.display_balance() ;
55
56   return 0 ;
57 }
```

The class constructor `bank_account()` *is automatically called, assigning the account number a value of* `123` *and the balance a value of* `10.54`

Line 44 of this program creates an object called `my_account` with these values:

my_account	
account_no	balance
123	10.54

Running the program will produce the following output:

```
Balance in Account 123 is 10.54
Balance in Account 123 is 21.04
Balance in Account 123 is 1.00
```

8.9 Default class constructor（默认的类构造函数）

Of course there is always the danger that the initial values in line 44 of program P8B are omitted. To allow for such a case, a default constructor can be included in the class. A default class constructor is a constructor that has no parameters.

Rewriting the bank account class to include a default class constructor:

当然，还存在这样一种隐患：忽略了程序P8B第44行中的初值。考虑到这种情况，可以在类中再包含一个默认构造函数，即不含参数的构造函数。

Program Example P8C

```
1  // Program Example P8C
2  // Demonstration of a class default constructor.
3  #include <iostream>
4  #include <iomanip>
5  using namespace std ;
6
7  class bank_account
8  {
9  public:
10     bank_account() ;
11     bank_account( int acc_no, double initial_balance ) ;
12     void deposit( double amount ) ;
13     void withdraw( double amount ) ;
14     void display_balance() ;
15  private:
16     int account_number ;
17     double balance ;
18  } ;
19
20  bank_account::bank_account()
21  {
22    account_number = 0 ;
23    balance = 0.0 ;
24  }
25
26  bank_account::bank_account( int acc_no, double initial_balance )
27  {
28    account_number = acc_no ;
29    balance = initial_balance ;
30  }
31
32  void bank_account::deposit( double amount )
33  {
34    balance += amount ;
35  }
36
```

The default class constructor has no parameters.

```
37 void bank_account::withdraw( double amount )
38 {
39    balance -= amount ;
40 }
41
42 void bank_account::display_balance()
43 {
44    cout << "Balance in Account " << account_number << " is "
45         << fixed << setprecision( 2 )
46         << balance << endl ;
47 }
48
```

Using the default constructor in main ():

```
49 int main( void )
50 {
51    bank_account my_account ;          ◄──────
52
53    my_account.display_balance() ;
54
55    my_account.deposit( 10.50 ) ;
56
57    my_account.display_balance() ;
58
59    my_account.withdraw( 20.04 ) ;
60
61    my_account.display_balance() ;
62
63    return 0 ;
64 }
```

The default class constructor bank_account() *is automatically called assigning the account number a value of* 0 *and the balance a value of* 0.0

Line 51 creates the object my_account with these values:

my_account	
account_no	balance
0	0.0

Running this program will display the following:

```
Balance in Account 0 is 0.00
Balance in Account 0 is 10.50
Balance in Account 0 is -9.54
```

8.10   Overloading class constructors
（重载类构造函数）

The bank_account class of program P8C has now got two constructors. One of the constructors does not have any parameters

程序P8C中的bank_account 类含有两个构造函数。其中一个构造

(the default constructor) and the other constructor has two parameters. There can be several constructors within a class, each with a different number of parameters and so, like any other function, constructors can be overloaded.

The next program extends the bank_account class by adding another constructor that has just one parameter. This parameter is used for setting the value of the bank account number.

函数没有任何参数（默认的构造函数），另一个构造函数有两个参数。

一个类中可以有若干个参数个数不同的构造函数。因此，构造函数和其他函数一样，也可以被重载。

Program Example P8D

```cpp
1  // Program Example P8D
2  // Demonstration of overloaded class constructors.
3  #include <iostream>
4  #include <iomanip>
5  using namespace std ;
6
7  class bank_account
8  {
9  public:
10   bank_account() ;
11   bank_account( int acc_no ) ;
12   bank_account( int acc_no, double initial_balance ) ;
13   void deposit( double amount ) ;
14   void withdraw( double amount ) ;
15   void display_balance( void ) ;
16 private:
17   int account_number ;
18   double balance ;
19 } ;
20
21 bank_account::bank_account()
22 {
23   account_number = 0 ;
24   balance = 0.0 ;
25 }
26
27 bank_account::bank_account( int acc_no )
28 {
29   account_number = acc_no ;
30   balance = 0.0 ;
31 }
32
33 bank_account::bank_account( int acc_no, double initial_balance )
34 {
35   account_number = acc_no ;
36   balance = initial_balance ;
37 }
38
39 void bank_account::deposit( double amount )
40 {
```

```
41    balance += amount ;
42  }
43
44  void bank_account::withdraw( double amount )
45  {
46    balance -= amount ;
47  }
48
49  void bank_account::display_balance( void )
50  {
51    cout << "Balance in Account " << account_number << " is "
52          << fixed << setprecision( 2 )
53          << balance << endl ;
54  }
55
56  int main( void )
57  {
58    bank_account my_account( 123 ) ;
59
60    my_account.display_balance() ;
61
62    return 0 ;
63  }
```

Because there is only one argument given when the object my_
account is being created, the class constructor on lines 27 to 31 is
called, assigning 123 to the account number and 0.0 to the balance.

因为在创建my_account对象时只
提供了一个实参，所以第27行至
第31行之间的构造函数被调用，
将账号和余额分别赋值为123和
0.0。

my_account	
account_no	balance
123	0.0

8.11 Constructor initialisation lists
（构造函数初始化列表）

The data members, account_number and balance, of the
bank_account class are initialised using assignment statements,
e.g. lines 35 and 36.

A data member initialisation list is frequently used in constructors in
place of assignment statements. For example, the constructor on lines
33 to 37 can be re-written as:

在构造函数中，经常用数据成员
初始化列表来代替赋值语句。

```
bank_account::
bank_account( int acc_no, float initial_balance ) :
              account_number( acc_no ), balance( initial_balance )
{
}
```

This is an example of a constructor initialisation list. In a constructor initialisation list, a colon follows the constructor function header. The data member `account_number` is initialised with the value of `acc_no` and the data member `balance` is initialised with the value of `initial_balance`.

In this case, the initialisation list performs all the initialisations required of the constructor. Hence, there are no statements left in this constructor. Note, however, that the chain brackets {} are still required.

在这个构造函数初始化列表中，构造函数头部后面有一个冒号。数据成员account_number用acc_no来初始化，数据成员balance 用initial_balance 来初始化。

在这种情况下，初始化列表执行了构造函数所需的全部初始化工作。所以，在构造函数里就不需要再写什么语句了。但是要注意，花括号{}不能省略。

8.12　Default argument values in a constructor
（构造函数中的默认实参值）

In chapter 7, default argument values were used in standalone nonclass functions. Since constructors are also functions, default arguments can also be used in constructors.

Instead of using the two overloaded constructors in lines 27 to 37 of program P8D, the data member `balance` can be assigned a default value of `0.0`. This is demonstrated in the next program:

在第7章不涉及类的独立的函数中，使用过默认的实参值。既然构造函数也是函数，那么默认的实参值也同样可以在构造函数中使用。

Program Example P8E

```
1  // Program Example P8E
2  // Demonstration of default arguments in a class constructor.
3  #include <iostream>
4  #include <iomanip>
5  using namespace std ;
6
7  class bank_account
8  {
9  public:
10   bank_account() ;
11   bank_account( int acc_no, double initial_balance = 0.0 ) ;
12   void deposit( double amount ) ;
13   void withdraw( double amount ) ;
14   void display_balance( void ) ;
15  private:
16   int account_number ;
17   double balance ;
18  } ;
19
20 bank_account::bank_account()
21 {
22   account_number = 0 ;
23   balance = 0.0 ;
24 }
25
26 bank_account::
```

Default argument in a constructor.

```
27 bank_account( int acc_no, double initial_balance ) :
28          account_number( acc_no ), balance( initial_balance )
29 {}              Can use an initialisation list and assignment statements in a constructor.
30
31 void bank_account::deposit( double amount )
32 {
33   balance += amount ;
34 }
35
36 void bank_account::withdraw( double amount )
37 {
38    balance -= amount ;
39 }
40
41 void bank_account::display_balance( void )
42 {
43    cout << "Balance in Account " << account_number << " is "
44        << fixed << setprecision( 2 )
45        << balance << endl ;
46 }
47
48 int main( void )
49 {
50   bank_account account1 ( 1 ) ;
51   // The constructor starting on line 26 is called,
52   // assigning the account number a value of 1 and the balance
53   // the default value of 0.0
54
55   bank_account account2 ( 2, 10.55 ) ;
56   // The constructor starting on line 26 is again called,
57   // assigning the account number a value of 2 and the balance
58   // a value of 10.55
59
60   bank_account account3 ;
61   // The default constructor on lines 20 to 24 is called,
62   // assigning the account number a value of 0 and the balance
63   // a value of 0.00
64
65   account1.display_balance() ;
66   account2.display_balance() ;
67   account3.display_balance() ;
68
69   return 0 ;
70 }
```

The output from this program is:

```
Balance in Account 1 is 0.00
Balance in Account 2 is 10.55
Balance in Account 0 is 0.00
```

8.13　**static class data members**
　　（静态类数据成员）

A static class data member is independent of all the objects that are created from that class. Only one copy of a static data member exists and is shared by all objects of a particular class. The value of the static data member is therefore the same for all the class objects. If even one of the objects of a class modifies the value of a static data member, then the value of the static data member changes for every object of that class.

静态数据成员独立于从类中创建的所有对象。静态数据成员对于类的所有对象只有一个副本，被该类的所有对象共享。所以，对于该类的所有对象，静态数据成员的值都是相同的。如果该类的一个对象修改了静态数据成员的值，那么该类的每个对象的这个静态数据成员的值也都随之改变。

The next program uses a static data member next_account_number that is incremented every time a bank account object is created. Using next_account_number, it is possible to automatically assign a bank account number to each new object of the class.

Program Example P8F

```
1  // Program Example P8F
2  // Demonstration of a class static data member.
3  #include <iostream>
4  #include <iomanip>
5  using namespace std ;
6
7  class bank_account
8  {
9  public:
10   bank_account() ;
11   bank_account( int acc_no ) ;
12   bank_account( int acc_no, double initial_balance ) ;
13   void deposit( double amount ) ;
14   void withdraw( double amount ) ;
15   void display_balance( void ) ;
16  private:
17   static int next_account_number ;
18   int account_number ;
19   double balance ;
20  } ;
21
22  bank_account::bank_account()
23  {
24   account_number = next_account_number++ ;
25   balance = 0.0 ;
26  }
27
28  bank_account::bank_account( int acc_no )
29  {
30   account_number = acc_no ;
31   balance = 0.0 ;
32  }
33
```

static class data member.

```
34  bank_account::bank_account( int acc_no, double initial_balance )
35  {
36    account_number = acc_no ;
37    balance = initial_balance ;
38  }
39
40  void bank_account::deposit( double amount )
41  {
42    balance += amount ;
43  }
44
45  void bank_account::withdraw( double amount )
46  {
47    balance -= amount ;
48  }
49
50  void bank_account::display_balance( void )
51  {
52    cout << "Balance in Account " << account_number << " is "
53         << fixed << setprecision( 2 )
54         << balance << endl ;
55  }
56
57  int bank_account::next_account_number = 1 ;
58
59  int main( void )
60  {
61    bank_account account1, account2, account3 ;
62
63    account1.deposit( 25.50 ) ;
64    account2.deposit( 30.50 ) ;
65    account3.deposit( 10.00 ) ;
66    account1.withdraw( 20.04 ) ;
67
68    account1.display_balance() ;
69    account2.display_balance() ;
70    account3.display_balance() ;
71
72    return 0 ;
73  }
```

A static data member is initialised outside the class and outside main().

The output from this program is:

```
Balance in Account 1 is 5.46
Balance in Account 2 is 30.50
Balance in Account 3 is 10.00
```

The new static data member next_account_number is declared on line 17. Since a static data member is independent of objects of the class, it must be assigned a value outside the class. This is done on line 57, before main() is entered. The scope resolution operator :: is

由于静态数据成员独立于类的所有对象，因此必须在类的外面对它进行赋值。

used to specify that `next_account_number` belongs to the class `bank_account`.

The class data member `next_account_number` occurs only once and is shared by all objects of the class, regardless of the number of bank account objects that are created. In contrast, the variables `account_number` and `balance` exist for every bank account object that is created.

类的数据成员`next_account_number`仅仅出现一次，无论创建多少个银行账户类对象，该类的所有对象都可以共享访问这个类的数据成员。相反地，变量`account_number`和`balance`则仅属于每个创建的银行账户类对象。

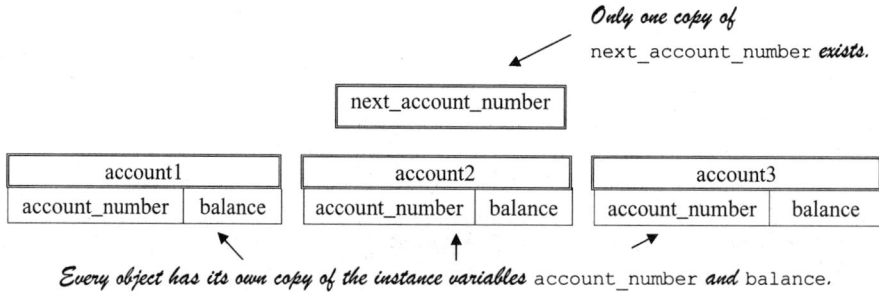

Only one copy of `next_account_number` *exists.*

next_account_number

account1		account2		account3	
account_number	balance	account_number	balance	account_number	balance

Every object has its own copy of the instance variables `account_number` *and* `balance`.

Because `next_account_number` is associated with a class rather than an object, `next_account_number` is called a *class variable*. The variables `account_no` and `balance` are associated with instances of the class and are called *instance variables*.

Line 61 defines three instances of the `bank_account` class: `account1`, `account2` and `account3`. The default constructor is called for all three objects.

由于静态数据成员`next_account_number`和类相关联而不是和对象相关联，因此称之为类变量。变量`account_no`和`balance`都与类的实例相关联，所以称之为实例变量。

- The account number for `account1` is assigned the initial value of `next_account_number` (which is 1) on line 24. After this assignment, `next_account_number` is incremented to 2.
- When the constructor for `account2` is called the value of `next_account_number` has remained at 2, because `account_number` is `static` and holds its value between calls to the constructor. Line 24 assigns the data member `account_number` of `account2` the value of 2 and increments `next_account_no` to 3.
- When the constructor for `account3` is called, the data member `account_number` for this object is assigned the value of 3.

In this way, each instance of the `bank_account` class is automatically assigned an account number.

8.14   Using `return` statement in a member function
（在成员函数中使用return语句）

The member functions of the bank account class do not have a `return` statement in them and so are type `void`. A class member function can, like any function, return a value using a `return` statement.

The next program calculates the total of the balances in all accounts. To do this a member function `get_balance()` is added to the `bank_account` class. This function simply returns the value of the `private` class data member balance for a bank account object.

类成员函数也可以像其他函数一样，使用return语句返回一个值。

Program Example P8G

```
1  // Program Example P8G
2  // Demonstration of a class function that returns a value.
3  #include <iostream>
4  #include <iomanip>
5  using namespace std ;
6
7  class bank_account
8  {
9  public:
10   bank_account() ;
11   bank_account( int acc_no ) ;
12   bank_account( int acc_no, double initial_balance ) ;
13   void deposit( double amount ) ;
14   void withdraw( double amount ) ;
15   void display_balance( void ) ;
16   double get_balance( void ) ;
17  private:
18   static int next_account_number ;
19   int account_number ;
20   double balance ;
21  } ;
22
23  bank_account::bank_account()
24  {
25   account_number = next_account_number++ ;
26   balance = 0.0 ;
27  }
28
29  bank_account::bank_account( int acc_no )
30  {
31   account_number = acc_no ;
32   balance = 0.0 ;
33  }
34
35  bank_account::bank_account( int acc_no, double initial_balance )
36  {
```

```
37    account_number = acc_no ;
38    balance = initial_balance ;
39  }
40
41  void bank_account::deposit( double amount )
42  {
43    balance += amount ;
44  }
45
46  void bank_account::withdraw( double amount )
47  {
48      balance -= amount ;
49  }
50
51  void bank_account::display_balance( void )
52  {
53    cout << "Balance in Account " << account_number << " is "
54         << fixed << setprecision( 2 )
55         << balance << endl ;
56  }
57
58  double bank_account::get_balance( void )
59  {
60      return balance ;
61  }
62
63  int bank_account::next_account_number = 1 ;
64
65  int main( void )
66  {
67    // Create four accounts.
68    bank_account account1, account2, account3, account4 ;
69
70    // Put some money into each account.
71    account1.deposit( 125.55 ) ;
72    account2.deposit( 130.75 ) ;
73    account3.deposit( 100.25 ) ;
74    account4.deposit( 300.45 ) ;
75
76    // Display the account numbers and the balances.
77    account1.display_balance();
78    account2.display_balance();
79    account3.display_balance();
80    account4.display_balance();
81
82    // Calculate the total amount on deposit.
83    float total_balances = account1.get_balance() +
84                           account2.get_balance() +
85                           account3.get_balance() +
86                           account4.get_balance() ;
```

```
87
88   cout << "Total Balances = " << fixed << setprecision( 2 )
89        << total_balances << endl ;
90   return 0 ;
91 }
```

The output from this program is:

```
Balance in Account 1 is 125.55
Balance in Account 2 is 130.75
Balance in Account 3 is 100.25
Balance in Account 4 is 300.45
Total Balances = 657.00
```

The member function get_balance() is called an *inspector* or *accessor* function. Inspector functions allow private data members of a class to be inspected from outside the class.

The class member functions deposit() and withdraw() are examples of *mutator* functions. Mutator functions change the values of the private data members of a class.

成员函数get_balance( )称为检查或读值函数。读值函数允许在类外读取类的私有数据成员的值。像deposit( )和withdraw( )这样的成员函数称为设值函数。设值函数用来修改类的私有数据成员的值。

8.15   Inline class member functions
（内联成员函数）

Using a member function like get_balance() simply to get the value of a data member of an object is a bit costly in terms of the overheads involved in calling a function. Inline functions are useful in cases like this where you want to use a function but don't want the function overheads.

Like the standalone non-class functions of chapter 7, class member functions can be either inline or non-inline. The member functions of a class can be made inline by preceding the function header with the keyword inline. For example, to make get_balance() inline, change line 16 of program P8G to:

```
inline double get_balance() ;
```

An alternative way of making a member function inline is to include the function definition in the body of the class. For example,

在使用像get_balance( )这样的成员函数来简单地读取一个对象的数据成员的值时，将增大函数调用方面的程序开销。如果想使用一个函数但又不想增大函数调用的开销，那么采用内联函数是一种解决方案。和第7章不涉及类的那些独立的函数一样，类成员函数可以是内联的，也可以是非内联的。在类的成员函数的函数头前面加上关键字inline，就可以把这个函数指定成为内联函数。把一个成员函数指定为内联函数的另一种方法是把函数定义放在类的内部。

```
class bank_account
{
public:
  bank_account() ;
  bank_account( int acc_no ) ;
  bank_account( int acc_no, double initial_balance ) ;
  void deposit( double amount ) ;
  void withdraw( double amount ) ;
  void display_balance( void ) ;
  double get_balance( void )
```

```
    {
        return balance ;
    }
    private:
        static int next_account_number ;
        int account_number ;
        double balance ;
} ;
```

inline is implicit for functions defined within the class.

A disadvantage of making functions inline by moving their definitions into the class is that the class becomes 'crowded' with details. A programmer who wants to use the class will want to know what functions are in the class and how to use them; details of how member functions are implemented may be of no interest.

这种把函数定义移至类的内部，使其成为内联函数的方法的缺点是：它会使类变得很"拥挤"。使用这个类的程序员只想知道这个类有哪些函数及如何使用它们，而对于这些成员函数是如何实现的可能并不感兴趣。

8.16 Class interface and class implementation
（类的接口和类的实现）

The public interface of a class is a list of the member functions of the class and how they can be used. The public interface starts at `public` and ends at the keyword `private`.

A programmer using a class should only have to read the public interface in order to use the class; knowledge of the subsequent details in the `private` section and in the member functions should not be necessary. These details are called the *class implementation* and are of concern only to the programmer who wrote or maintains the class.

To make the public interface easier to read, comments should be included explaining the purpose of the functions, their parameters and their return values. For example:

类的公共接口是一个成员函数的列表及它们的使用方法，从关键字public开始，到关键字private结束。

程序员要使用这个类，只需了解这个公共接口，无须关注类的私有部分和成员函数的内部实现细节，这些细节称为类的实现。只有负责编写或者维护这个类的程序员才会关注它。

```
1   class bank_account
2   {
3   public:
4   // Constructors.                    Public interface of the class.
5     bank_account() ;
6   // Purpose: Default class constructor.
7   //          First instance of the class is assigned account number
8   //          1, the second instance account number 2, and so on.
9   //          The account balance is set to 0 for each instance.
10
11    bank_account( int acc_no ) ;
12  // Purpose  : Class constructor to set the account number
13  //            of an instance of the class to the parameter value.
14  // Parameter: An account number.
15
16    bank_account( int acc_no, double initial_balance ) ;
17  // Purpose  : Constructor to set the account number and balance of a
18  //            class instance to specified values.
19  // Parameters: An account number and a balance.
```

```
20
21 // Mutator functions.
22   void deposit( double amount ) ;
23   // Purpose  : Function to add an amount to the account balance.
24   // Parameter: Amount of money.
25
26   void withdraw( double amount ) ;
27   // Purpose  : Function to subtract an amount from the balance.
28   // Parameter: Amount of money.
29
30 // Inspector functions.
31   void display_balance( void ) ;
32   // Purpose: Displays the account balance.
33
34   inline double get_balance( void ) ;
35   // Purpose: To inspect the account balance of a class instance.
36   // Returns: Private class member account_balance.
37 private:
38   static int next_account_number ;
39   int account_number ;
40   double balance ;
41 } ;
42
43 bank_account::bank_account()
44 {
45   account_number = next_account_number ++ ;
46   balance = 0.0 ;
47 }
48
49 bank_account::bank_account( int acc_no )
50 {
51   account_number = acc_no ;
52   balance = 0.0 ;
53 }
54
55 bank_account::bank_account( int acc_no, double initial_balance )
56 {
57   account_number = acc_no ;
58   balance = initial_balance ;
59 }
60
61 void bank_account::deposit( double amount )
62 {
63   balance += amount ;
64 }
65
66 void bank_account::withdraw( double amount )
67 {
68   balance -= amount ;
69 }
70
71 void bank_account::display_balance( void )
```

Class implementation details include the private *section and the* public *member functions.*

```
72 {
73    cout << "Balance in Account " << account_number << " is "
74         << fixed << setprecision( 2 )
75         << balance << endl ;
76 }
77
78 double bank_account::get_balance( void )
79 {
80    return balance ;
81 }
82
83 int bank_account::next_account_number = 1 ;
```

8.16.1 Separation of class interface and class implementation

Separating the public interface of a class from its implementation makes sense for a programmer who is just interested in how to use the class. Unfortunately, it is not possible in C++ to completely separate the public interface of a class from its implementation. In the bank account class, for example, the public interface and the implementation details of the private data members must be enclosed together between braces { and }. In C++, the class declaration is normally placed in a *header* file (e.g. `bank_ac.h`) and the member functions are normally placed in a separate file (e.g. `bank_ac.cpp`).

For example, placing line 1 to 41 of the bank account class declaration in a file `bank_ac.h`:

对于只关心如何使用这个类的程序员而言，把类的公共接口和类的实现分离开来是很有意义的。遗憾的是，在C++中把公共接口与类的实现完全分离是不可能的。例如，在银行账户类中，公共接口和私有数据成员的实现必须放在一对花括号中。

在C++中，类的声明通常放在一个头文件（例如bank_ac.h）中，成员函数的定义则放在另一个单独的文件（例如bank_ac.cpp）中。

```
// Declaration of bank_account class.

#if !defined BANK_AC_H
#define BANK_AC_H

class bank_account
{
public:
// Constructors.
  bank_account() ;
  // Purpose: Default class constructor.
  //          First instance of the class is assigned account number
  //          1, the second instance account number 2, and so on.
  //          The account balance is set to 0 for each instance.
...
private:
  static int next_account_number ;
  int account_number ;
  double balance ;
} ;


#endif
```

The lines beginning with # are standard preprocessor directives used to prevent multiple inclusions of the header file into a program. See appendix G for details.

以#号开头的行是标准预处理命令，用于防止在一个程序中多次包含同一个头文件，详见附录G。

Placing lines 43 to 83 of the bank account class into `bank_ac.cpp`:

```cpp
// Member function definitions and static data member
initialisation
// code for bank account class.
#include <iostream>
#include <iomanip>
#include "bank_ac.h"
using namespace std ;

bank_account::bank_account()
{
  account_number = next_account_number ++ ;
  balance = 0.0 ;
}
...
int bank_account::next_account_number = 1 ;
```

Using these two files, program example P8G can be re-written as:

Program Example P8H

```cpp
1   // Program Example P8H
2   // Demonstration of using class header and source code files.
3   #include <iostream>
4   #include <iomanip>
5   #include "bank_ac.h"
6   #include "bank_ac.cpp"
7   using namespace std ;
8
9   int main( void )
10  {
11    // Create four accounts.
12    bank_account account1, account2, account3, account4 ;
13
14    // Put some money into each account.
15    account1.deposit( 125.55 ) ;
16    account2.deposit( 130.75 ) ;
17    account3.deposit( 100.25 ) ;
18    account4.deposit( 300.45 ) ;
19
20    // Display the account numbers and the balances.
21    account1.display_balance();
22    account2.display_balance();
23    account3.display_balance();
24    account4.display_balance();
```

```
25
26   // Calculate the total amount on deposit.
27   float total_balances = account1.get_balance() +
28                          account2.get_balance() +
29                          account3.get_balance() +
30                          account4.get_balance() ;
31
32   cout << "Total Balances = " << fixed << setprecision( 2 )
33        << total_balances << endl ;
34   return 0 ;
35 }
```

Lines 5 and 6 are preprocessor directives that incorporate the bank account interface and implementation files into the program. The quotes around the file names on lines 5 and 6 tell the compiler that the files to be included are in the same directory as the program. See appendix G for details.

第5行和第6行是预处理命令，用来把银行账户文件的接口和实现文件合并到本程序中。第5行和第6行中文件名两端的双引号的作用是告诉编译器要包含的文件和本程序位于同一目录下，详见附录G。

8.16.2　Use of namespaces in header files

A program may contain many classes and functions #included from many different header files. These header files may have been written by different programmers and may contain many elements such as classes, functions and objects. An element in one of the header files may have inadvertently been given the same name as an element in another header file. This will result in a compiler error.

一个程序可能包含来自不同头文件的类和函数。这些头文件可能是由不同的程序员编写的，可能包含诸如类、函数和对象等各种元素。稍不小心，就会在不同的头文件中出现元素同名的情况，从而引起编译错误。

C++ uses *namespaces* to resolve this problem. A namespace is a named block of statements in a program. Within a namespace, an identifier name must be unique although the name may be previously used in other namespaces.

C++使用命名空间来解决这个问题。命名空间就是程序中一个已命名的语句块。在一个命名空间内，标识符的名字必须是唯一的，尽管这个名字先前可能在其他命名空间中使用了。

As an example, suppose alpha is one of many classes defined in a class library header file classlib1.h containing the following:

```
#if !defined CLASS_LIB1_H
#define CLASS_LIB1_H

namespace classlib1
{
  class alpha
  {
  private:
    int x, y ;
  public:
    ...
  } ;
  class beta
  {
    ...
  } ;
```

```
   ...
}
```

```
#endif
```

Another class library header file, `classlib2.h`, contains a different class with the same name `alpha`.

```
#if !defined CLASS_LIB2_H
#define CLASS_LIB2_H

namespace classlib2
{
  class alpha
  {
  private:
    int z ;
  public:
    ...
  } ;
    ...
}
```

```
#endif
```

There are two ways to distinguish between the two versions of `alpha` in a program. The first way is to use the scope resolution operator (`::`).

有两种方法可以区分这两个同名的类。第一种方法是使用作用域运算符（`::`）。

```
#include "classlib1.h"
#include "classlib2.h"
#include <iostream>
using namespace std ;

int main( void )
{
  classlib1::alpha o1 ; // o1 is a classlib1 alpha object.
  classlib2::alpha o2 ; // o2 is a classlib2 alpha object.
  ...
  return 0 ;
}
```

The second way to distinguish between the two versions of `alpha` is with the `using` statement. This statement indicates to the compiler which version of `alpha` to use.

第二种方法是使用using语句。这条语句会告诉编译器具体使用哪一个alpha类。

```
#include "classlib1.h"
#include "classlib2.h"
#include <iostream>
using namespace std ;

int main( void )
{
  using classlib1::alpha ;
```

```
  alpha o1 ; // o1 is a classlib1 alpha object.
  alpha o2 ; // o2 is also a classlib1 alpha object.
  ...
  return 0 ;
}
```

This form of the `using` statement provides access to a specific element in a namespace. Access to all the elements of a namespace can be achieved with a second form of the `using` statement:

```
#include "classlib1.h"
#include "classlib2.h"
#include <iostream>
using namespace std ;

int main( void )
{
  using namespace classlib1 ;
  // Can access all elements in the classlib1 namespace.
  alpha o1 ; // o1 is a classlib1 alpha object.
  alpha o2 ; // o2 is also a classlib1 alpha object.
  beta o3 ; // o3 is a classlib1 beta object.
  ...
  return 0 ;
}
```

The `std` namespace is used throughout the examples in this book. The standard library classes, functions and objects like `cin` and `cout` are defined in this namespace. It is not necessary to use the entire `std` namespace in a program. For example, if a program is just using `cout`, then instead of the statement

```
using namespace std ;
```

the statement

```
using std::cout ;
```

can be used.

使用这种格式的using语句可以访问一个命名空间中的一个特定的元素。要实现对一个命名空间中的所有元素的访问，需要使用using语句的第二种格式。

本书中的所有例子都使用std这个命名空间。像cin和cout等标准库类、函数和对象都是在这个命名空间中定义的。当然，没有必要在程序中使用全部的std命名空间。

Programming pitfalls

1. Don't forget to place a semi-colon after the last } in a class declaration.
2. A constructor has no return type, not even `void`.
3. A class constructor is called automatically when an object of the class is created, it cannot be called explicitly.
4. Do not include parentheses when a default constructor is used to create an object of a class. For example,

```
bank_account my_account() ; // Incorrect.
bank_account my_account ;   // Correct.
```

5. Non-inline functions must use the scope resolution operator (`::`).
6. `static` class data members must be initialised before the start of `main()`.
7. The default access of members in a class is private. Don't forget to place `public:` before member functions that are intended to be public.

1. 在类的声明中不要漏掉最后的花括号后面的分号。
2. 构造函数无返回类型，甚至也不能加void。
3. 创建类的对象时，这个类的构造函数会被自动地调用，不能显式地调用构造函数。
4. 使用默认构造函数创建类的对象时，不能包含圆括号。
5. 非内联函数必须使用作用域运算符(`::`)。
6. 静态类数据成员必须在main( )函数之前被初始化。
7. 在类中，成员的默认访问权限是私有的。如果要把一个成员函数指定为公有的，不要忘了在它前面加上"public:"。

Quick syntax reference

	Syntax	Examples

Declaring a class	`class class_name`
	`{`
	  `public:`
	  `// Public data members and`
	  `// functions.`
	  `private:`
	  `// Private data members and`
	  `// functions.`
	`} ;`
		`class bank_account`
		`{`
		  `public:`
		    `void display_balance() ;`
		    `...`
		  `private:`
		    `int account_number ;`
		    `float balance ;`
		    `static int`
		      `next_account_number ;`
		    `...`
		  `} ;`

Creating an instance of a class	`class_name variable1, variable2,`
	    `...                      ;`
		`bank_account ac1,`
		      `ac2,`
		      `ac3 ;`

Accessing class members	Member selection operator . (Dot operator)	`ac1.display_balance() ;`

Initialising a static class data member	`data type class name::`
	    `variable = a constant value ;`
		`int bank_account::`
		    `next_account_number = 1 ;`

Exercises

1. List the class data members and class member functions for
   (a) a digital alarm clock
   (b) a CD player
   (c) an elevator.

2. Briefly answer each of the following:

　(a) What is another name for a class data member?

　(b) What is the name given to a class member function that returns the value of a class data member?

　(c) How many parameters in a default constructor?

　(d) What is the public interface of a class?

　(e) What is the name given to a class member function that modifies the value of a class data member?

　(f) What is another name for a class member function?

　(g) Can a constructor return a value? Explain.

　(h) What is the purpose of the scope resolution operator ∷?

3. Find the errors in the following:

(a)
```
class class_a
{
private
  int a ;
public::
  class_a( int a_value ) ;
}
```

(b)
```
class class_b
{
  private:
  int b1;
  int b2 = 0 ;
  class_b() ;
} ;
```

4. Write a class declaration for each of the following classes. Include the member functions `assign_data()` to assign values to the data members and `display_data()` to display the values of the data members.

　(a) A class `current_date` with integer data members `day`, `month`, and `year`.

　(b) A class `current_time` with integer data members `hours`, `minutes`, and floating point data member `seconds`.

　(c) A class `complex` with floating point data members named `real` and `imaginary`.

　(d) A class `circle` with integer data members named `centre_x` and `centre_y` and a floating point data member `radius`.

　(e) A class `rectangle` with `double` data members `length` and `width`.

　(f) A class `cube` with an `unsigned` integer data member `size`.

　Create an object of each of the above classes in `main()`.

　Use `assign_data()` to assign values to the data members of each object.

　Display these values on the screen using `display_data()`.

5. Add a default constructor to each class in exercise 4.

6. Modify the classes developed in exercise 4 as follows:

　(a) Add a member function `increment_date()` to the `current_date` class that adds one day to the current date.

　(b) Add a member function `increment_time(s)` to the `current_time` class that adds s seconds to the current time.

　(c) Add a member function `calculate_magnitude()` to the `complex` class that calculates and returns the magnitude of a complex number.

　　(Note: the magnitude of a complex number a + bi is got by calculating the square root of $(a^2 + b^2)$).

　(d) Add a member function `area()` to the `circle` class that calculates and returns the area of a circle.

(e) Add a member function `perimeter()` to the `rectangle` class that calculates and returns the perimeter of a rectangle.

(f) Add a member function `volume()` to the `cube` class that calculates and returns the volume of a cube.

7. Modify program P8H to include a `static` class data member `overdrawn_fee` which is added to overdrawn accounts. Assume the overdrawn fee is 10 Yuan.

8. Create an elevator class for an elevator in a building with ten floors.

(a) The elevator starts at the first floor.

(b) When + is pressed the elevator goes to the next floor. Ignore this command if the elevator is on the tenth floor.

(c) When-is pressed the elevator goes to the previous floor. Ignore this command if the elevator is on the first floor.

(d) When S (shut down) is pressed the elevator should return to the first floor.

(e) As the elevator moves from floor to floor, the floor number should be displayed along with the bell (ASCII code 7) sounding.

Test the class in `main()` by creating an elevator object and continually inputting a character from the keyboard until an S (shut down) is entered.

9. The following is a class for recording the position of a motorised robot.

```
class robot
{
public:
  // Constructor with default arguments and constructor list.
  robot(float x = 0, float y = 0) : x_coord( x ), y_coord(y)
  {}
  // Inspector function to display the robot's position.
  void display_position()
  {
    cout << "(" << x_coord << "," << y_coord << ")" << endl ;
  }
private:
  float x_coord, y_coord ;
} ;
```

The following is a demonstration of the class `robot`:

```
int main( void )
{
  robot r2d2( 10.0, 8.1 ) ; // Constructor sets the initial
                            // position.

  r2d2.left( 1.3 ) ; // Move robot left 1.3 cms.
                  // New position is (8.7,8.1)
  r2d2.display_position() ;

  r2d2.back(4.21) ; // Move robot back 4.21 cms.
                  // New position is (8.7,12.31)
```

```
    r2d2.display_position() ;

    r2d2.right( 3.1 ) ; // Move robot right 3.1 cms.
                        // New position is (11.8,12.31)
    r2d2.display_position() ;

    r2d2.return_to_base() ; // Sets the position to (0,0).

    r2d2.forward( 0.3 ) // Move robot forward 3.1 cms.
                        // New position is (0,0.3).
    r2d2.display_position() ;

    r2d2.goto( 1.5, 4.5 ) ; // New position is (1.5,4.5).

    r2d2.return_to_base() ; // Move to position (0,0).
    return 0;
}
```

Write the member functions `left()`, `right()`, `forward()`, `back()`, `goto()` and `return_to_base()`.

10. Extend exercise 9 to include the following features:

(a) Don't allow the robot to crash into the boundary walls. The corners of the four boundary walls are at the x-y positions (0,0), (0,100), (100,100) and (100,0). These positions will be `static` data members of the class.

If the robot is going to crash into a wall with the next move, it should stop at 0.1 cm from the wall.

(b) Extend the program to allow a speed and a time to be specified. For example, if the robot is at (10,5) then `speed_left(4,2)` will result in the robot moving left at a speed of 4 cm/s for 2 seconds. The robot's new position will then be (2,5).

(c) The robot is fuelled with Xenotoplartogenicplasma (X for short). The robot travels 2000 cm per ml of X. If the robot is directed to a position that is so far away from its base that it cannot get back with the remaining fuel, then it should refuse to go.

(d) Use `return` statements in each member function that moves a robot to return to one of the following codes:

0 = move was successful

1 = move failure - fuel shortage

2 = move failure - move is outside a boundary wall

`main()` should now display an appropriate message every time the robot is moved.

Chapter Nine
Pointers and Dynamic Memory
第 9 章　指针和动态内存分配

9.1　**Variable addresses**（变量的地址）

Every variable and object used in a C++ program is stored in a specific place in memory. Each location in memory has a unique address, in the same way that every house in a street has a unique address.

The next program uses & to get the address of a variable and display it on the screen.

在C++程序中使用的每个变量和对象，都存储在内存中特定的存储单元中。每个存储单元都有唯一的地址，就像街道旁的每个房子都有唯一的地址一样。

Program Example P9A

```
1   // Program Example P9A
2   // Program to display the address of variables.
3   #include <iostream>
4   #include <iomanip>
5   using namespace std ;
6
7   int main( void )
8   {
9     int var1 = 1 ;
10    float var2 = 2 ;
11
12    cout << "var1 has a value of " << var1
13         << " and is stored at " << &var1 << endl ;
14    cout << "var2 has a value of " << var2
15         << " and is stored at " << &var2 << endl ;
16    return 0 ;
17  }
```

A sample run of this program is:

```
var1 has a value of 1 and is stored at 0012FF88
var2 has a value of 2 and is stored at 0012FF84
```

This is how the variables var1 and var2 are stored in memory:

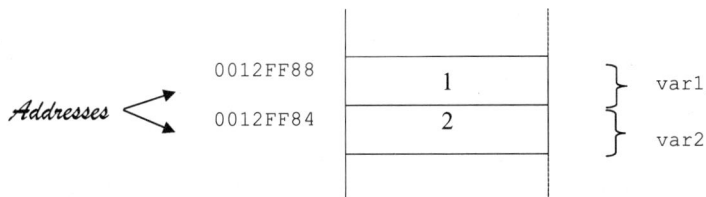

Different computers may give different addresses from the ones above. This is because various computers and operating systems will store variables at different memory locations. The addresses are in hexadecimal (base 16).

在不同的计算机上运行上面的程序时，输出的地址可能是不同的。这是因为不同的计算机和操作系统为变量分配的内存地址是不同的。地址都用十六进制数表示。

9.2　Pointer variables（指针变量）

A *pointer* variable is a variable that holds the address of another variable. A pointer variable is defined as follows:

指针变量是存放另一变量地址的变量。

```
data_type* variable_name ;
```

where `data_type` is any data type (such as `char`, `int`, `float`, a `struct`, a `class` and so on) and `variable_name` is any valid variable name. For example:

```
int* int_ptr ;     // int_ptr is a pointer to an int variable.
float* float_ptr ; // float_ptr is a pointer to a float variable.
bank_account* b ;  // b is a pointer to a bank_account object.
```

Whitespace in a pointer definition is not relevant. The pointer variable `int_ptr1`, for example, could also be defined as:

```
int * int_ptr ;
```

or

```
int *int_ptr ;
```

or

```
int*int_ptr ;
```

Pointer definitions are read backwards from the variable name, replacing * with the words "is a pointer". Thus, `int* int_ptr` means that `int_ptr` is a pointer to an `int`, and `float* float_ptr` means that `float_ptr` is a pointer to a `float`.

可以从变量名开始按照从后向前的顺序来读指针定义，并将"*"读作"是一个指针"。因此，"int*int_ptr"读作"int_ptr是一个指向整型变量的指针"，"float* float_ptr"读作"float_ptr是一个指向浮点型变量的指针"。

The next program defines and uses two pointer variables.

Program Example P9B

```
1   // Program Example P9B
2   // Demonstration of pointer variables.
3   #include <iostream>
4   using namespace std ;
5
6   int main( void )
7   {
8     int var1 = 1 ;
9     float var2 = 2 ;
10    int* ptr1 ;
11    float* ptr2 ;
12
13    ptr1 = &var1 ; // ptr1 contains the address of var1.
```

```
14   ptr2 = &var2 ; // ptr2 contains the address of var2.
15   cout << "ptr1 contains " << ptr1 << endl ;
16   cout << "ptr2 contains " << ptr2 << endl ;
17   return 0 ;
18 }
```

The output from this program is:

```
ptr1 contains 0012FF88
ptr2 contains 0012FF84
```

This is how the program variables are stored in memory:

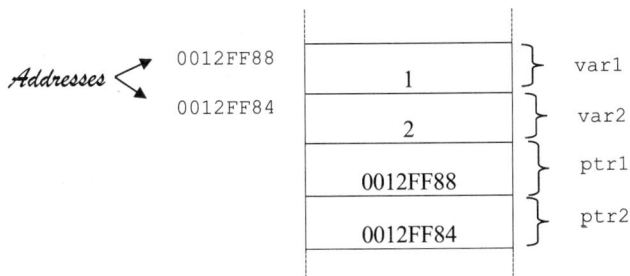

The two variables ptr1 and ptr2 are used to store the addresses of the other two variables, var1 and var2.

9.3   The dereference operator * （解引用运算符 * ）

The dereference operator * is used to access the value of a variable, whose address is stored in a pointer. For instance, *ptr means the value of the variable at the address stored in the pointer variable ptr. The dereference operator * is also called the indirection operator.

解引用运算符*用于访问指针变量所指向的存储单元中的内容。例如，*ptr为指针变量ptr所指向的存储单元中的内容。
解引用运算符*也称为间接寻址运算符。

Program Example P9C

```
1  // Program Example P9C
2  // Demonstration of dereference operator *
3  #include <iostream>
4  using namespace std ;
5
6  int main( void )
7  {
8    int var = 1 ;
9    int* ptr ;
10
11   ptr = &var ; // ptr contains the address of var
12   cout << "ptr contains " << ptr << endl ;
13   cout << "*ptr contains " << *ptr << endl ;
14   return 0 ;
15 }
```

The output from this program will be similar to the following:

```
ptr contains 0012FF88
*ptr contains 1
```

在上面的程序中，星号（*）出现了两次。第9行的*用来定义一个指向整型变量的指针ptr。第13行的*用来访问指针变量ptr所指向的存储单元中的内容。二者的用法是无关的。

The asterisk (*) is used in two different contexts in the above program. In line 9, the * is used to define `ptr` as a pointer to an `int`. In line 13, the * is used to access the value of the memory location, the address of which is in `ptr`. The two uses of * are not related.

程序的第11行将变量var的地址赋值给指针变量ptr。第12行显示存放在指针变量ptr中的地址。第13行利用解引用运算符*显示指针变量ptr所指向的存储单元中的内容，这称为指针变量ptr的解引用。

Line 11 of this program assigns the address of the variable `var` to the pointer variable `ptr`.

Line 12 displays the address contained in the pointer `ptr`.

Line 13 displays the value at the address held in ptr by using the dereference operator *. This is called *dereferencing* the pointer `ptr`. The value of `*ptr` is the same as the value of `var`.

当然，*ptr 与var 的值是相同的。

9.4　Using `const` with pointers
（使用const 修饰指针变量）

When defining a pointer, the pointer itself, the value it points to or both can be made constant. The position of `const` in the definition determines which of these apply.

定义指针变量时，指针变量本身、指针变量所指向的数据都可以声明为常量。究竟是谁被声明为常量，由变量定义中const的位置来决定。

In each of the following cases, assume two integer variables `i` and `j` are defined as:

```
int i, j ;
```

(a)

```
const int* p = &i ; // *p is a constant but p is not.
```

在下面几种情况中，都假设已经定义了两个整型变量i和j。

If a pointer definition is read backwards, what is being defined as a constant can be seen. The above definition reads as "p is a pointer to an integer constant". This means that the integer is constant and cannot be changed using pointer p with a statement such as:

```
*p = 5 ; // Illegal: cannot change i using p.
```

Note, however, the value of `i` can be changed with a statement such as

```
i = 5 ; // Legal: i is not a constant.
```

The pointer may be changed, so the following statement is legal:

```
p = &j ; // Legal: p now points to j.
```

如果按照从后向前的顺序来读取指针定义，那么哪一个被定义为常量就显而易见了。(a)中的定义语句可以读作"p是一个指向整型常量的指针变量"。这意味着p指向的整型数据是常量，不能使用指针变量p来修改它所指向的整型数据。但是可以直接修改i的值，也可以修改指针变量本身的值。

(b)

```
int const* p = &i ; // *p is a constant; p is not.
```

This definition of p reads as "p is a pointer to a constant integer". Again, this means that the integer is constant and cannot be changed using pointer p. This is an equivalent definition of p in (a) above.

(b)中的定义语句可以读作"p是一个指向常量整型的指针"。这也意味着p指向的整型数据是常量，它的值是不能使用指针变量p来修改的。(b)与(a)是等价的。

(c)

```
int* const p = &i ; // p is a constant; *p is not.
```

This definition of p reads, "p is a constant pointer to an integer". This means that the pointer is a constant but not what it points to.

```
*p = 5 ; // Legal: *p can be changed.
p = &j ; // Illegal: p is a constant.
```

(d)

```
const int* const p = &i ; // Both p and *p are constants.
```

This definition of p reads, "p is a constant pointer to an integer constant". This means that both the pointer and the integer it points to are constants.

```
*p = 5 ; // Illegal: *p is a constant.
p = &j ; // Illegal: p is a constant.
```

(c)中的定义语句可以读作"p是一个指向整型变量的常量指针"。这意味着指针变量本身是一个常量，而它指向的对象不是常量。

(d)中的定义语句可以读作"p是一个指向整型常量的常量指针"。这意味着指针变量本身和指针变量所指向的整型数据都是常量。

9.5 Pointers and one-dimensional arrays
（指针和一维数组）

Pointers and arrays are directly related to one another. In C++, the name of an array is equivalent to the address of the first element of the array. The name of an array, therefore, is a pointer to the first element of the array. Consider the following array definition:

```
int a[5] ;
```

The elements of this array are: a[0], a[1], a[2], a[3], and a[4]. The name of the array is a, and this is equivalent to the address of the first element; in other words, a is the same as &a[0]. The following program demonstrates this.

指针和数组是相互关联的。在C++中，数组名代表数组的首地址。因此，数组名即为指向数组第一个元素的指针。

a为数组名，也是数组的首地址，也就是说，a等价于&a[0]。

Program Example P9D

```
1   // Program Example P9D
2   // Program to show that the name of an array is the same
3   // as the address of its first element.
4   #include <iostream>
5   using namespace std ;
6
7   int main( void )
8   {
9     int a[5] ;
10
11    cout << "a is " << a
12        << " and &a[0] is " << &a[0] << endl ;
13    return 0 ;
14  }
```

The output from this program is:

```
a is 0012FF78 and &a[0] is 0012FF78
```

The actual addresses may be different on your system, but the two addresses will be the same nonetheless.

虽然程序实际输出的地址值因使用的操作系统的不同而有所不同，但是不管怎样，a和&a[0]这两个地址值一定是相同的。

正如a代表数组的第一个元素的

Just as a is the address of the first element, a + 1 is the address of the second element, a + 2 is the address of the third element, and so on. As the name of an array is a pointer to the first element of the array, the dereference operator * can be used to access the elements of the array. The next program demonstrates a commonly used technique of displaying the elements of an array using the dereference operator *.

地址一样，a + 1代表数组的第二个元素的地址，a + 2代表数组的第三个元素的地址，以此类推。由于数组名为指向数组的第一个元素的指针，因此可以使用解引用运算符*来访问数组元素的值。

Program Example P9E

```
1   // Program Example P9E
2   // Program to access the elements of an array by using
3   // element addresses rather than subscripts.
4   #include <iostream>
5   using namespace std ;
6
7   int main( void )
8   {
9     int a[5] = { 10, 13, 15, 11, 6 } ;
10
11    for ( int i = 0 ; i < 5 ; i++ )
12      cout << "Element " << i << " is " << *( a + i ) << endl ;
13    return 0 ;
14  }
```

This program displays the elements of the array as follows:

```
Element 0 is 10
Element 1 is 13
Element 2 is 15
Element 3 is 11
Element 4 is 6
```

If *(a + i) in line 12 is changed to a[i], the program would produce the same output. Thus:

```
*( a + 0 ) or *a is equivalent to a[0]
*( a + 1 )        is equivalent to a[1]
*( a + 2 )        is equivalent to a[2], and so on.
```

The parentheses in the expression *(a + i) are important. Without them the expression *a + i would add the first element of the array a and i together.

You can use pointers to access the elements of any array, not just an array of integers. If an array of floats is defined as:

```
float numbers[100] ;
```

then numbers[i] is equivalent to *(numbers + i).

Although the name of an array is a pointer to the first element of the array, you cannot change its value; this is because it is a constant pointer. Expressions such as a++ or numbers+=2 are invalid, because both a and numbers are array names. You can, however, assign the name

表达式*(a+i)中的圆括号是非常重要的。如果去掉圆括号，表达式*a+i就表示数组a的第一个元素的值与i的值相加。

可以使用指针来访问任意数组的元素，而不仅仅是整型数组。

虽然数组名是指向数组第一个元素的指针，但由于它是一个常量指针，因此它的值是不能改变的；因为a和numbers都是数组名，所以像a++或numbers+=2这样的表达式是不合法的。但是，

of an array to a pointer variable of the same type. For example:

可以将数组名赋值给一个类型相同的指针变量。

```
int a[5] ;
int* p ;
p = a ;      // Valid: assignment of a constant to a variable.
a++ ;        // Invalid: the value of a constant cannot change.
p++ ;        // Valid: p is a variable. p now points to
             // element 1 of the array a.
p-- ;        // Valid: p points to element 0 of the array a.
p += 10 ;    // Valid, but p is outside the range of the array a,
             // so *p is undefined. A common error.
p = a - 1 ; // Valid, but p is outside the range of the array.
```

A constant may be added to or subtracted from the value of a pointer, allowing access to different memory locations. However, not all arithmetic operations are permissible on pointers. For example, the multiplication of two pointers is illegal, because the result would not be a valid memory address.

通过将指针变量加上或减去一个常量，可以使它指向不同的存储单元。但是，并非所有的算术运算符都可以应用于指针。例如，两个指针相乘是非法的，因为相乘的结果可能不是一个有效的内存地址。

9.6   Pointers and multi-dimensional arrays
（指针和多维数组）

As with one-dimensional arrays, you can access the elements of a multi-dimensional array using pointers. However, as the number of dimensions of an array increases, the pointer notation becomes increasingly complex. Consider the following definition of a two-dimensional array a:

像一维数组一样，也可以利用指针来访问多维数组的元素。但是，随着数组维数的增加，指针的表示方法将更加复杂。

```
int a[3][2] = { { 4, 6 },
                { 1, 3 },
                { 9, 7 } } ;
```

A two-dimensional array is stored as an "array of arrays". This means that a is a one-dimensional array whose elements are themselves a one-dimensional arrays of integers.

As with a one-dimensional array, the name of a two-dimensional array is a pointer to the first element of the array. Therefore a is equivalent to &a[0]. a[0] is itself an array of two integers, which means that a[0] is equivalent to &a[0][0].

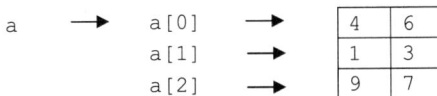

在C++中，二维数组以"数组的数组"的形式存储，这意味着a是一个一维整型数组，它的元素又是一个一维整型数组。

与一维数组一样，一个二维数组的数组名也是一个指向数组第一个元素的指针。因此，a等价于&a[0]。而a[0]本身又是一个具有两个整型元素的一维数组，即a[0]等价于&a[0][0]。

```
a   ⟶   a[0]   ⟶   ┌───┬───┐
                    │ 4 │ 6 │
        a[1]   ⟶   ├───┼───┤
                    │ 1 │ 3 │
        a[2]   ⟶   ├───┼───┤
                    │ 9 │ 7 │
                    └───┴───┘
```

a[0], a[1] and a[2] are pointers (data type is int*) and a is a pointer to a pointer (data type is int **).

a[0] is the address of the first element in the first row of the array.

　　*a[0] is a[0][0], which is 4.

a[1] is the address of the first element in the second row.

a[0]、a[1]和a[2]是指针（数据类型为int*），a 是一个指向指针的指针（数据类型为int **）。

a[0]代表数组a第1行第一个元素的地址。

a[1]代表数组a第2行第一个元素的地址。

`*a[1]` is `a[1][0]`, which is 1.

a[2] is the address of the first element in the third row.

   `*a[2]` is `a[2][0]`, which is 9.

a[2]代表数组a第3行第一个元素的地址。

a[0]+1 is the address of the second element in the first row.

   `*(a[0]+1)` is `a[0][1]`, which is 6.

a[0]+1代表数组a第1行第二个元素的地址。

a[1]+1 is the address of the second element in the second row.

   `*(a[1]+1)` is `a[1][1]`, which is 3.

a[1]+1代表数组a第2行第二个元素的地址。

a[2]+1 is the address of the second element in the third row.

   `*(a[2]+1)` is `a[2][1]`, which is 7.

a[2]+1代表数组a第3行第二个元素的地址。

Using the fact that

$$*a \text{ is the same as } a[0]$$
$$\text{and } *(a + 1) \text{ is the same as } a[1]$$
$$\text{and } *(a + 2) \text{ is the same as } a[2]$$

the following can be derived:

```
1. a[0][0] is *a[0] is *(*a) or **a
2. a[1][0] is *a[1] is *(*(a+1))
3. a[2][0] is *a[2] is *(*(a+2))
4. a[0][1] is *(a[0]+1) is *(*a+1)
5. a[1][1] is *(a[1]+1) is *(*(a+1)+1)
6. a[2][1] is *(a[2]+1) is *(*(a+2)+1)
```

9.7　Pointers to structures（指向结构体的指针）

In addition to defining a pointer to a variable of a built-in data type, it is also possible to define a pointer to a variable of a type defined by `struct` or `class`.

不仅可以定义指针，使其指向一个具有内置数据类型的变量，还可以定义指针，使其指向一个 struct或者class类型的变量。

The general format for defining a pointer to a structure is:

```
struct tag_name* variable_name ;
```

where `tag_name` is the structure tag and `variable_name` is the name of the pointer variable. For example, consider the `student_rec` structure used in chapter 5.

```
struct student_rec // Structure template.
{
  int number ;
  float scores[5] ;
} ;

struct student_rec student ; // Define a structure variable.
```

The following line defines a pointer `ptr` to the `student_rec` structure:

```
struct student_rec *ptr ;
```

A value can be assigned to `ptr` by using the address operator &, as in:

```
ptr = &student ;
```

可以使用取地址运算符&给ptr赋值。

Note that it is the address of the structure variable `student` and not the address of the structure tag `student_rec` that is assigned to `ptr`.

注意这里赋给ptr的值是结构体变量student的地址，而不是结构体标记student_rec的地址。

The members of a structure variable can be referenced by using the dereference operator *. For example,

结构体变量的成员可以用解引用运算符*来访问。

```
(*ptr).number
```

will access the student's number. The parentheses are necessary, because the selection operator . has a higher priority than the dereference operator *. Without the parentheses `*ptr.number` is attempting to access the memory location given by `ptr.number`. This is invalid, because `ptr` is not a structure and `number` is not a member of `ptr`. C++ provides a much more convenient notation for accessing the members of a structure. The arrow notation `->` (– and > together) can be used in place of the dot notation. Thus,

C++提供了更加方便的符号来访问结构体的成员，即可以用指向运算符"->"（-和>结合）来代替成员选择运算符"."。

```
ptr -> number     and      (*ptr).number
```

are equivalent. The expression `ptr->number` reads as "the member `number` of the structure pointed to `by ptr`".

9.8 **Pointers to class objects**
（指向类对象的指针）

Defining a pointer to a class object is similar to defining a pointer to a structure variable.

定义一个指向类对象的指针与定义一个指向结构体变量的指针类似。

The general format for defining a pointer to a class object is:

```
class_name* variable_name ;
```

where `class_name` is the name of the class and `variable_name` is the name of the pointer variable.

The next program demonstrates the use of a pointer to an object of the bank account class of program P8H.

Program P9F

```
1  // Program P9F
2  // Demonstration of a pointer to a class object.
3  #include <iostream>
4  #include <iomanip>
5  #include "bank_ac.h"
6  #include "bank_ac.cpp"
```

```
7   using namespace std ;
8
9   int main( void )
10  {
11    bank_account ac ;        // ac is a bank_account object.
12    bank_account* ac_ptr ;  // ac_ptr is a pointer to a bank_account.
13
14    ac_ptr = &ac ;  // ac_ptr contains the address of the object ac.
15    ac_ptr -> deposit( 100 ) ;
16    ac_ptr -> display_balance() ;
17    return 0 ;
18  }
```

The output from this program is:

```
Balance in account 1 is 100.00
```

Line 12 defines ac_ptr as a pointer to a bank_account object and line 14 assigns the address of the bank_account object ac to ac_ptr.

The public members of a class object may be accessed by using the dereference operator *. For example,

类对象的公有成员可以通过解引用运算符*来访问。

```
(*ac_ptr).deposit( 100 )
```

will call the public member function deposit().

The first pair of parentheses are necessary, because the selection operator. has a higher priority than the dereference operator *.
As with structures, the arrow notation -> can be used in place of the dot notation. Thus,

与解引用运算符*相比，成员选择运算符.具有更高的优先级。与结构体类似，也可以用指向运算符->来代替成员选择运算符.。二者相比，->更方便、更常用。

```
ac_ptr -> deposit( 100 ) ;    and    (*ac_ptr).deposit( 100 ) ;
```

are equivalent. Of the two notations, -> is more convenient and common.

9.9   Pointers as function arguments
（指针变量作为函数实参）

Like any data type, pointers can be used as function arguments. The next program is a re-write of program P7H, which used reference variables to swap two values. This version of the program uses pointers in place of references.

与其他数据类型一样，指针变量也可以作为函数的实参。在程序P7H中利用引用变量实现了交换两个变量的值，下面的程序利用指针变量代替引用变量对其进行了改写。

Program Example P9G

```
1   // Program Example P9G
2   // Demonstration of pointer arguments.
3   #include <iostream>
4   using namespace std ;
5
6   void swap_vals( float* val1, float* val2 ) ;
7   // Purpose   : To swap the values of two float variables.
8   // Parameters: Pointers to the two float variables.
```

```
9
10 int main( void )
11 {
12    float num1, num2 ;
13
14    cout << "Please enter two numbers: " ;
15    cin >> num1 ;
16    cin >> num2 ;
17    // Swap values around so that the smallest is in num1
18    if ( num1 > num2 )
19      swap_vals( &num1, &num2 ) ;
20    cout << "The numbers in order are "
21         << num1 << " and " << num2 << endl ;
22    return 0 ;
23 }
24
25 void swap_vals( float* ptr1, float* ptr2 )
26 {
27    float temp = *ptr1 ;
28
29    *ptr1 = *ptr2 ;
30    *ptr2 = temp ;
31 }
```

Pointer arguments.

A sample run of this program is:

```
Please enter two numbers: 12.1  6.4
The numbers in order are 6.4 and 12.1
```

Line 19 passes the addresses of the two floating-point variables num1 and num2 to the function swap_vals(). These addresses are received by the parameters ptr1 and ptr2, declared as pointers to floats in the function header on line 25.

Line 27 stores the value of num1 (= *ptr1) in the variable temp (temp is now 12.1).

The statement

```
*ptr1 = *ptr2 ;
```

in line 29 is equivalent to

```
num1 = num2 ;
```

because *ptr1 is the same as num1 and *ptr2 is the same as num2.

Therefore, num1 gets the value 6.4.

Finally, the statement

```
*ptr2 = temp ;
```

in line 30 assigns the value of temp (12.1) to num2, because *ptr2 is the same as num2.

The result is that the values in num1 and num2 are swapped.

将上面程序与程序P7H进行对比，可以看出使用引用变量比使用指针变量更简单。在程序的第19行

Comparing this program with program P7H, it can be seen that it is easier to use references rather than pointers. When calling the function on line 19, & must be used to pass the address of the variables to the function. Also, within the function the dereference operator * must be used to access the value of each of the numbers. Forgetting to do these two operations is a common error when using pointers as function arguments.

Although references are more convenient to use, pointers are important because of the number of library functions that use them as parameters. One such function is ctime(), which converts the time in seconds to a character string containing the date and time.

进行函数调用时，必须使用取地址运算符&将变量的地址传递给函数，而且在函数体内部，必须使用解引用运算符*来访问每个数据值。一个常见错误是，使用指针变量作为函数实参时，忘记使用这两个运算符。

虽然使用引用变量更方便，但是指针也是非常重要的，因为很多库函数都是用指针变量作为形参，例如用于将以秒为单位的时间转换成包括日期和时间在内的字符串的库函数ctime( )。

Program Example P9H

```
1   // Program Example P9H
2   // Demonstration of ctime() library function.
3   #include <iostream>
4   #include <ctime>
5   #include<string>
6   using namespace std ;
7
8   int main( void )
9   {
10    time_t current_time ; // Define a variable of type time_t.
11
12    current_time = time( 0 ) ; // Get the current time in seconds.
13    // Display the current date and time as a text string.
14    cout << "Current date and time: " << ctime( &current_time ) << endl ;
15    return 0 ;
16  }
```

A sample run of this program is:

```
Current date and time: Wed Feb 11 08:47:52 2009
```

More common examples involve arrays as function parameters. For example, line 21 of program P7I is commonly written as:

```
int sum_array( int* array, int no_of_elements )
```

The first argument on line 16 of P7I is the name of an array, which also a pointer to the first element of the array. Therefore, the first parameter on line 21 is a pointer to an integer and can also be written as int*array.

9.10　Dynamic memory allocation（动态内存分配）

When defining an array, the number of elements in the array must be specified in advance of the program execution. Sometimes, either all the elements specified are not used or more elements than were originally anticipated are required. To avoid these problems, C++ has

定义一个数组时，必须在程序执行前指定数组元素的个数。有时，要么是已指定的数组元素都没有被使用，要么是实际需要的数组元素的个数超出了最初预期

the ability to allocate memory while a program is executing. This is done using the memory allocation operator `new`.

9.10.1   Allocating memory dynamically for an array

The `new` memory allocation operator can be used to allocate a contiguous block of memory for an array of any data type, whether the data type is built-in or is a user-defined structure or class.

The general format of `new` for allocating memory for an array is:

```
pointer = new data_type[ size ] ;
```

where `pointer` is a pointer to the start of the allocated memory, `data_type` is the data type of the array and `size` is the number of elements in the array. For example:

```
// Allocate memory for 10 integers.
int_ptr = new int[10] ;

// Allocate memory for 5 bank account objects.
ac_ptr = new bank_account[5]
```

When allocating memory for an array of class objects, there must be a default constructor for the class so that the elements of the array get initialised.

The next program demonstrates `new` by allowing the user to specify the number of elements in an integer array while the program is running.

的个数。为了避免这些问题的发生，C++ 允许在程序执行过程中动态地分配内存，可以通过使用内存分配运算符new来实现。动态内存分配运算符new可以为任何类型的数组分配一个连续的存储空间，无论该数组类型是内置数据类型还是用户自定义的结构体或类。

这里pointer是指向动态分配的内存首地址的指针，data_type是数组数据类型，size是数组中元素的个数。

当为一个类对象数组动态分配内存时，这个类必须有自己的默认构造函数，以初始化类对象数组的元素。

Program Example P9I

```
1   // Program Example P9I
2   // Demonstration of dynamic memory allocation for an array
3   // of integers.
4   #include <iostream>
5   using namespace std ;
6
7   int main( void )
8   {
9     int* int_array ;
10    int no_els, i ;
11
12    cout << "Enter the number of elements " ;
13    cin >> no_els ;
14    // Allocate the required memory while the program is running.
15    int_array = new int[no_els] ;
16    // Enter the elements into the array.
17    for ( i = 0 ; i < no_els ; i++ )
18    {
19      cout << "Enter element " << i << ": " ;
20      cin >> int_array[i] ;
21    }
22    // Display the element values just entered.
```

```
23   for ( i = 0 ; i < no_els ; i++ )
24     cout << "Element " << i << " is " << *( int_array + i ) << endl ;
25   delete[] int_array ; // Free the allocated memory.
26   return 0 ;
27 }
```

A sample run of this program is:

```
Enter the number of elements 2
Enter element 0 57
Enter element 1 69
Element 0 is 57
Element 1 is 69
```

The number of array elements required is input from the keyboard on line 13. Line 15 uses new to allocate the exact number of elements required. The operator new stores the starting address of the allocated memory block in the pointer int_array.

The elements of the newly allocated array can be accessed using either a pointer or an index. For example, int_array[0] and *int_array both access the first element of the array and int_array[1] and *(int_array + 1) both access the second element of the array. The index notation is used on line 20 and the pointer notation is used on line 24.

可以通过指针或下标方式访问新分配内存的数组元素。

The allocated memory is freed using the delete operator. It is important to remember to include the square brackets [] when freeing memory for a previously allocated array. Without the square brackets, only the first element of the array will be deleted.

可以使用delete运算符释放已分配的内存。切记：当为一个数组释放先前动态分配的内存时，必须使用方括号[ ]。如果不加方括号，那么释放的仅是数组的第一个元素的内存。

9.10.2 Initialisation with new

When allocating memory for an array of class objects, the default constructor for the class initialises the elements of the array. For example,

当为一个类对象的数组动态分配内存时，会自动调用类的默认构造函数对数组元素进行初始化。

```
ac_ptr = new bank_account[5] ;
```

results in the default constructor for the bank account class being called five times.

No initialisation is done for dynamically allocated arrays of built-in data types. For example,

为内置数据类型的数组动态分配内存时无须对其进行初始化。

```
int_ptr = new int[10] ;
```

results in the ten non-initialised integer elements.

Single instances of any data type (built-in, a user-defined structure or a class) can be initialised using a second form of the new operator. The general format of new for a single instance of a data type is:

对于任意数据类型（内置数据类型、用户自定义结构体或类）的单个实例，可以使用new运算符的另一种形式进行初始化。

```
pointer = new data_type( initial_value ) ;
```

where pointer is a pointer to the allocated memory. The initial value is optional.

```
// Allocate memory for an integer, with an initial value of 100.
int_ptr = new int( 100 ) ;

// Allocate memory for an integer, with no initial value.
int_ptr = new int ;

// Allocate memory for a bank account object. The default
// class constructor is called to do the initialisation.
ac_ptr = new bank_account ;

// Allocate memory for a bank account object. A constructor
// is called to assign initial values to account number
// and balance.
ac_ptr = new bank_account( 1234, 100 ) ;
```

The next program demonstrates the use of new with various
initialisations.

Program Example P9J

```
1   // Program Example P9J
2   // Demonstration of initialisation with the operator new.
3   #include <iostream>
4   #include <iomanip>
5   #include "bank_ac.h"    // See 8.16.1
6   #include "bank_ac.cpp"
7
8   using namespace std ;
9
10  int main( void )
11  {
12    int no_of_acs ;
13
14    // Dynamically create an array of bank accounts.
15    // The default constructor is called for each array element.
16    cout << "Enter the number of bank accounts " ;
17    cin >> no_of_acs ;
18    bank_account* accounts = new bank_account[no_of_acs] ;
19    // Display the initialised elements of the array.
20    cout << endl << "Accounts:" << endl ;
21    for ( int i = 0 ; i < no_of_acs ; i++ )
22        accounts[i].display_balance() ;
23    cout << endl ;
24
25    // Create a single instance of a bank account.
26    // The default constructor is called to do the initialisation.
27    cout << "bank_ptr1:" << endl ;
28    bank_account* bank_ptr1 = new bank_account ;
29    bank_ptr1->display_balance() ;
30    cout << endl ;
31
32    // Create a single instance of a bank account.
33    // Initialisation is done by the third constructor in the class.
34    bank_account* bank_ptr2 = new bank_account( 123, 100 ) ;
```

```
35    cout << "bank_ptr2:" << endl ;
36    bank_ptr2->display_balance() ;
37    delete[] accounts ;
38    delete bank_ptr1 ;
39    delete bank_ptr2 ;
40    return 0 ;
41 }
```

A sample run of this program is:

```
Enter the number of bank accounts 2

Accounts:
Balance in Account 1 is 0.00
Balance in Account 2 is 0.00

bank_ptr1:
Balance in Account 3 is 0.00

bank_ptr2:
Balance in Account 123 is 100.00
```

9.10.3   Allocating memory for multi-dimensional arrays

In C++, multi-dimensional arrays are implemented as 'arrays of arrays'. To fully understand dynamic memory allocation for multi-dimensional arrays, familiarity with section 9.6 is necessary.

The next program demonstrates the dynamic allocation of a two-dimensional array of integers.

在C++中，多维数组按照"数组的数组"的形式来实现。为了更好地理解如何为多维数组进行动态内存分配，有必要先熟悉一下9.6节的内容。

Program Example P9K

```
1  // Program Example P9K
2  // Dynamic allocation of a two-dimensional array.
3  #include <iostream>
4  #include <iomanip>
5  using namespace std ;
6
7  int main( void )
8  {
9    int no_of_rows, no_of_cols ;
10   int i, j ;
11   float **data ;
12
13   cout<< "Number of rows: " ;
14   cin >> no_of_rows ;
15   cout<< "Number of columns: " ;
16   cin >> no_of_cols ;
17
18   // Allocate requested storage:
19
20   // (a) allocate storage for the rows.
21   data = new float* [no_of_rows] ;
22
23   // (b) allocate storage for each column.
```

```
24   for ( j = 0 ; j < no_of_rows; j++ )
25     data[j] = new float[no_of_cols] ;
26
27   // Place some values in the array.
28   for ( i = 0 ; i < no_of_rows ; i++ )
29     for ( j = 0 ; j < no_of_cols ; j++ )
30       data[i][j] = i * 10 + j ;
31
32   // Display elements of the array.
33   for ( i = 0 ; i < no_of_rows ; i++ )
34   {
35     for ( j = 0 ; j < no_of_cols ; j++ )
36       cout << data[i][j] << ' ' ;
37     cout << endl ;
38   }
39
40   // Free the allocated storage:
41
42   // (a) delete the columns.
43   for ( i = 0 ; i < no_of_rows ; i++ )
44     delete[] data[i] ;
45
46   // (b) delete the rows.
47   delete[] data ;
48   return 0 ;
49 }
```

A sample run of this program is:

```
Number of rows: 3
Number of columns: 4
0 1 2 3
10 11 12 13
20 21 22 23
```

Lines 44 and 47 free the memory allocated in lines 21 and 25 separately. Note that for each pointer returned from new in lines 21 and 25 there is a corresponding call to delete with that pointer in lines 47 and 44.

程序的第44行和第47行分别释放了由第21行和第25行所分配的内存空间。注意：对于第21行和第25行中每一个由new运算符返回的指针，在第47行和第44行中相应地使用delete运算符进行释放。

9.10.4 Out of memory error

In the previous programs, it was assumed that the memory requested with new was allocated, regardless of whether memory was available or not. C++ handles insufficient memory errors produced by new by calling a function specified in set_new_handler().

The next program continually allocates memory in one megabyte blocks until no more memory exists. When new is unable to allocate memory, the function out_of_memory() is called.

前面的程序假设：无论内存是否够用，使用new运算符申请内存都能进行分配。C++ 通过调用函数set_new_handler()来 处 理new运算符引起的内存不足的错误。

Program Example P9L

```
1 // Program Example P9L
2 // Demonstration of error handling with the new operator.
```

```
3   #include <iostream>
4   using namespace std ;
5
6   void out_of_memory( void ) ;
7
8   int main( void )
9   {
10    const int ONE_MB = 1024 * 1024 ;
11    int memory_allocated = 0 ;
12    int* ptr ;
13
14    set_new_handler( out_of_memory ) ;
15
16    for ( ; ; ) // An infinite loop.
17    {
18      ptr = new int[ONE_MB] ; // Allocate memory in 1MB blocks.
19      memory_allocated++ ;
20      cout << memory_allocated << " MB allocated..." << endl ;
21    }
22    return 0 ;
23  }
24
25  void out_of_memory( void )
26  {
27    cerr << "Error: Out of memory" << endl ;
28    exit( 1 ) ;
29  }
```

A sample run of this program is:

```
1 MB allocated...

2 MB allocated...
...
419 MB allocated...
500 MB allocated...
Error: Out of memory
```

The function `out_of_memory()` inserts an error message into the stream `cerr` rather than into `cout`. The stream `cerr` is typically used for error messages while `cout` is used for displaying the results of a program. Like `cout`, `cerr` is, by default, connected to the screen. However, the output for `cout` is often redirected to a device other than the screen (e.g. a disk file). In this case `cout` is unsuitable for error messages that may require immediate attention, so `cerr` is used instead.

Redirecting output streams to different devices is done by the operating system commands, not by C++.

Line 28 terminates the program and exits to the operating system with a status code of 1. A non-zero status code is usually used to indicate an abnormal exit from a program.

函数out_of_memory( )将错误提示信息送到输出流cerr中而不是cout中。cerr主要用来处理错误信息，而cout通常用于显示程序的运行结果。默认情况下，cout与cerr都与屏幕相关联，但是cout通常会被重定向到屏幕以外的其他设备（如磁盘文件）。这时，cout就不适合显示错误信息了，因为错误信息需要立即显示以引起用户的注意。此时，应该使用cerr。将输出流重定向到哪个设备是由操作系统命令而非C++程序来指定的。

Programming pitfalls

1. Consider the following code segment:

```
int a = 1, b = 2, c ;
int* pa = &a, *pb = &b ;

c = *pa/*pb ;
```

The `/*` in the above assignment is interpreted as the start of a C-style comment. Use parentheses, as in:

```
c = (*pa)/(*pb) ;
```

or use spaces, as in:

```
c = *pa / *pb ;
```

2. When defining two or more pointers as it is a common error to write the definition as:

```
int* p1, p2 ;
```

However, this defines `p1` as a pointer to an integer and `p2` as an integer.

The correct definition is:

```
int *p1, *p2 ;
```

or

```
int* p1, *p2 ;
```

3. Pointers, like any other variable, are not automatically initialised. Do not use a pointer until it has been assigned a value. For example:

```
int* p ;
*p = 100 ; // Where is p pointing to?
```

4. The `new` operator uses `()` to initialise a single instance of a data type and `[]` to allocate memory for an array. For example:

```
int* p = new int(10) ; // Allocate memory for an int with an
                       // initial value of 10.
int* p = new int[10] ; // Allocate memory for an array of 10
                       // integers.
```

5. Use `delete` if an object was created with `new` without `[]` and use `delete[]` if the object was created with `new[]`.

6. Be careful not to create a *lingering* or *dangling* pointer, i.e. a pointer that points to a memory block that has been de-allocated with `delete`. For example:

```
int* p1 = new int ( 1 ) ;
int* p2 = p1 ;
delete p1 ;
cout << *p2 ; // *p2 no longer exists!
```

1. 思考下面的程序段存在什么错误。

上面赋值语句中的/*会被误认为C风格注释的开始，因此必须添加圆括号。

2. 定义两个或两个以上的指针变量时，写成如下定义形式是一个常见错误。

3. 与其他变量相同，指针变量不能自动初始化。在使用指针变量前，必须先对其赋初值。

4. new运算符使用( )是初始化一个数据类型的单个实例，使用[ ]则是为数组分配动态内存。

5. 如果一个对象是使用不带[ ]的new运算符创建的，那么必须使用delete释放为其分配的内存。如果一个对象是使用new[ ]创建的，那么必须使用delete[ ]释放为其分配的内存。

6. 注意不要产生悬挂指针，也就是指向一个已用delete释放的内存块的指针变量。

7. A *memory leak* can occur when an allocated block of memory has no associated pointer. For example:

7. 对于一块动态分配的内存，如果不再有指针变量指向它，那么将发生内存泄漏。

```
int* p = new int ( 1 ) ; // p points to a memory block
                         // with the number 1 in it.
p = new int ( 2 ) ; // p now points to the memory block
                    // with the number 2 in it.
                    // The first memory block is inaccessible.
```

Quick syntax reference

	Syntax	Examples

Defining a pointer	`data_type* variable ;`	`int* pa ;`
		`float* pb, *pc ;`
		`bank_account* bank_ac_ptr ;`

Address	`&variable`	`int a ;` `pa = &a ;`
Dereference *	`*variable`	`a = *pa ;`
The name of an array is a pointer to the first element of the array		`int a[10] ;` `// a is the same as &a[0].`

Memory allocation	`pointer = new [size] ;`
	or
	`pointer = new (value) ;`
		`// Allocate storage for`
		`// an array of integers.`
		`int* pi ;`
		`pi = new [10] ;`
		`// Allocate storage for a`
		`// float, initial value 10.`
		`float *pf ;`
		`pf = new float( 10 ) ;`

Free memory	`delete[] pointer ;`
	or
	`delete pointer ;`
		`delete[] pi ;`
		`delete pf ;`

Exercises

1. Write a program to define the following variables and to display their addresses:

```
char c = 'a' ;
int i = 54321 ;
short s = 123 ;
float f = 125.5 ;
double d = 1234.25 ;
```

Draw a diagram to illustrate the memory layout for these variables.

How many bytes of memory are allocated for each of these variables?

2. Given the following:

```
int* i_ptr ;
float* f_ptr ;
int i = 1, k = 2 ;
float f = 10.0 ;
```

which of these statements are valid?

(a) `i_ptr = &i ;`   (b) `f_ptr = &f ;`   (c) `f_ptr = f ;`
(d) `f_ptr = &i ;`   (e) `k = *i ;`   (f) `k = *i_ptr ;`
(g) `i_ptr = &k ;`   (h) `*i_ptr = 5 ;`   (i) `i_ptr = &5 ;`

3. What does this program segment display?

```
int a, b ;
int* p1, *p2 ;
a = 1 ;
b = 2 ;
p1 = &a ;
p2 = &b ;
b = *p1 ;
cout << a << ' ' << b << endl ;
cout << *p1 << ' ' << *p2 << endl ;
*p1 = 15 ;
cout << a << ' ' << b << endl ;
*p1 -= 3 ;
cout << a << ' ' << b << endl ;
*p2 = *p1 ;
cout << a << ' ' << b << endl ;
(*p1)++ ;
cout << a << ' ' << *p2 << endl ;
p1 = p2 ;
*p1 = 50 ;
cout << a << ' ' << b << endl ;
```

4. What is wrong with each of the following?

(a) `float a[5] ;`   (b) `float a[5] ;`   (c) `float a[5] ;`
    `int* p ;`   `float* p ;`   `float* p ;`
    `p = a ;`   `p = &a ;`   `p = a ;`

(d) `int n = new int ;`   (e) `int* p = new float( 5 ) ;`
    `n = 0 ;`   `p[0] = 0 ;`

5. What is the output from the following?

```
string* sp1 = new string ( "asdfghjk" ) ;
string* sp2 ;
string s = *sp1 ;
string& r = s ; // r is a reference to s.
sp2 = &s ;      // sp2 contains the address of s.
s.at( 0 ) = 'A' ;
sp1 -> erase ( 2, 3 ) ;
cout << s << endl ;
cout << r << endl ;
cout << *sp1 << endl ;
cout << *sp2 << endl ;
```

6. What does this program segment do?

```
int a[5] ;
int *p ;
for ( int i = 0 ; i < 5 ; i++ )
  cin >> *(a+i) ;
```

```
for ( p = a ; p < a+5 ; p++ )
   cout << ' ' << *p ;
```

7. What is the value of `*p`, `*p+4` and `* (p+4)` in each of the following?

(a) ```
int one_d[] = {1,3,4,5,-1} ;
int *p ;
p = one_d ;
```
(b) ```
float f[] = { 1.25, 11.0, 9.5, 3.5, 6.5, 1.0 } ;
float *p ;
p = f ;
```
(c) ```
int two_d[3][6] = { {1, 5, 0, 9,11, -4},
                    {3, 9, 4, 6, 10, 123},
                    {11, 7, 4, -10, 19, 15} } ;
int *p ;
p = two_d[1] ;
```

8. Given the following definitions:

```
int numbers[10] = { 1,7,8,2 } ;
int *ptr = numbers ;
```

what is in the array `numbers` after each of the following?

(a) `* (ptr+4) = 10 ;`
(b) `*ptr-- ;`
(c) `* (ptr+3) = * (ptr+9) ;`
(d) `ptr++ ;`
(e) `*ptr = 0 ;`
(f) `* (numbers+1) = 1 ;`

9. If `a` is a 6 by 8 array, which elements of `a` do the following expressions access?

(a) `*a[2]`
(b) `* (a[2]+7)`
(c) `* (*a)`
(d) `* (* (a+5)+2)`

10. Using `new`, write a program to input a specified number of integer values into an array and to display the array and the sum of the elements in the array. Use pointers, not subscripts, in the program.

11. Given an array such as

```
int a[] = { 1, 2, 9, -1, 4, 0, -2, 8 } ;
```

write a program to replace the elements with a value less than 0 with 0. Use pointers.

12. Given the following arrays:

```
float litres[] = { 11.5, 11.21, 12.7, 12.6, 12.4 } ;
float kilometres[] = { 186.5, 184.7, 231.5, 234.1, 232.6 } ;
int KPL[5] ; // Kilometres Per Litre
```

write a program to calculate and display the value of each element of `KPL`.
Use pointers to access the elements of each array.
```

Chapter Ten
Operator Overloading
第 10 章　运算符重载

10.1　The need for operator overloading
（运算符重载的必要性）

The C++ built-in arithmetic (+, *, / etc.) and relational operators
(>, <, ==, !=) work with built-in data types (int, float, etc.).
However, not all the built-in operators can be used with every data
type. For example, multiplication is not a valid operation for strings
and % can only be used with integer data types. Of course, it doesn't
make sense to multiply two strings together, so the * operator is
inappropriate for strings. The + operator, however, is appropriate
for strings and is taken to mean concatenation (joining one string of
characters to the end of another). This means that the operator + has
a different meaning for numeric data types than for string data types.

When you create a new class you can redefine or overload existing
operators such as + to work appropriately on objects created from that
class. For example, the addition of an integer to a date such as
31/12/1999 + 1 should result in the date 1/1/2000, not 32/12/1999 nor
31/12/2000. When using dates, the + operator does not mean simple
addition and will have to be *overloaded* to work in an appropriate
manner for dates.

Although it makes sense to add an integer to a date, it doesn't make
sense to add two dates together. However, the subtraction of two
dates is meaningful and results in the number of days between the
two dates. Other data types where different meanings are assigned to
+ and − are time (11:59 + 0:01 = 12:00) and angles (9°10'−20' = 8°−50').
Existing operators are overloaded for specific use in a class use by
writing *operator functions* for the class.

C++ 的内置数据类型（int，float，
等等）可与内置算术运算符（+，
*,/, 等等）和关系运算符（>,<，
==, !=）配合使用。但是，并非
所有的内置运算符都能与每一种
数据类型配合使用。例如，字符
串不能进行乘法操作，%只适用
于整数。当然，两个字符串相乘
是没有意义的，因此，运算符*不
能作用于字符串。但是，运算符
+可作用于字符串，它表示字符
串连接（将一个字符串连接到另
一个字符串的尾部）。这说明运
算符+作用于数值型数据和作用
于字符串时，其含义是不同的。
定义一个新的类时，可以重新定
义或者重载已经存在的运算符。

通过为类添加运算符函数来实现
对已有的运算符重载，从而完成
特定的操作。

10.2　Overloading the addition operator +
（重载加法运算符+）

The next program demonstrates the use of an overloaded operator by
including an addition operator + function for class objects that store
times in the twenty-four hour format.

Program Example P10A

```
1   // Program Example P10A
2   // Demonstration of an overloaded + operator.
3   #include <iostream>
4   using namespace std ;
5
6   class time24 // A simple 24-hour time class.
7   {
8   public:
9     time24( int h = 0, int m = 0, int s = 0 ) ;
10    void set_time( int h, int m, int s ) ;
11    void get_time( int& h, int& m, int& s ) const ;
12    time24 operator+( int s ) const ;
13  private:
14    int hours ;   // 0 to 23
15    int minutes ; // 0 to 59
16    int seconds ; // 0 to 59
17  } ;
18
19  // Constructor.
20  time24::time24( int h, int m, int s ) :
21            hours( h ), minutes( m ), seconds( s )
22  {}
23
24  // Mutator function.
25  void time24::set_time( int h, int m, int s )
26  {
27    hours = h ; minutes = m ; seconds = s ;
28  }
29
30  // Inspector function.
31  void time24::get_time( int& h, int& m, int& s ) const
32  {
33    h = hours ; m = minutes ; s = seconds ;
34  }
35
36  // Overloaded + operator.
37  time24 time24::operator+( int s ) const
38  {
39    // Add s to class member seconds and calculate the new time.
40    time24 temp ;
41    temp.seconds = seconds + s ;
42    temp.minutes = minutes + temp.seconds / 60 ;
43    temp.seconds %= 60 ;
44    temp.hours = hours + temp.minutes / 60 ;
45    temp.minutes %= 60 ;
46    temp.hours %= 24 ;
47    return temp ; // Return the new time.
48  }
49
```

```
50  int main( void )
51  {
52    int h, m, s ;
53    time24 t1( 23, 59, 57 ) ; // t1 represents 23:59:57
54    time24 t2 ;
55
56    t2 = t1 + 4 ;                // t2 should now be 0:0:1
57    t2.get_time ( h, m, s ) ;
58    cout << "Time t2 is " << h << ":" << m << ":" << s << endl ;
59    return 0 ;
60  }
```

The output from this program is:

```
Time t2 is 0:0:1
```

The prototype for the overloaded + operator is on line 12. This line specifies that an integer is passed to the operator+ function and the function returns a time24 object. The const at the end of the declaration ensures that the object that called the function is not modified within the function.

Lines 37 to 48 define the overloaded + operator for adding an integer number of seconds to a time24 object and returning the result. Like any other function, operator+() has a return type and a name: operator+. Since operator+() is like any other function, it can be called like any other function; line 56 can also be written as:

```
t2 = t1.operator+( 4 ) ;
```

The call to the operator + function on line 56 is obviously more intuitive, but the above form is useful in understanding how the function is actually called and the type of argument it takes.

The way the + operator has been overloaded in this program restricts its use to adding an int to a time24 object. An expression such as t3 = t1 + t2 is not valid, because this expression is equivalent to t3 = t1.operator+(t2). The problem here is that the parameter of the member function operator+ is an integer and t2 is a time24 object.

The solution to this problem is to write another operator+ member function that has a time24 object as a parameter. This is done in the next program.

第12行是重载运算符+的函数的函数原型。函数operator+以一个整型数据作为形参，返回一个类time24的对象，声明语句末尾的const确保调用该函数的对象在函数中不会被修改。

像其他函数一样，重载运算符函数operator+()也有其返回类型及函数名（operator+），可以像调用其他函数那样来调用它。

第56行对函数operator +的调用方式更直观，但上面的函数调用形式更有助于理解函数是如何调用的及它所需的实参的类型。

在此程序中，这种重载运算符+的方式将它的用途限定为一个整型数据与一个time24对象相加。因此，像t3 = t1 + t2这样的表达式就不是合法的，因为这个表达式等价于t3 = t1.operator + (t2)，出错的原因是成员函数operator+要求的形参是一个整型数据，而t2却是一个time24对象。要解决这个问题，可以为类time24再添加一个以time24对象为形参的operator+成员函数。

Program Example P10B

```
1  // Program Example P10B
2  // Demonstration of an overloaded + operator: version 2.
3  #include <iostream>
4  using namespace std ;
5
6  class time24  // A simple 24-hour time class.
```

```
7   {
8   public:
9       time24( int h = 0, int m = 0, int s = 0 ) ;
10      void set_time( int h, int m, int s ) ;
11      void get_time( int& h, int& m, int& s ) const ;
12      time24 operator+( int secs ) const ;
13      time24 operator+( const time24& t ) const ;
14  private:
15      int hours ;     // 0 to 23
16      int minutes ;   // 0 to 59
17      int seconds ;   // 0 to 59
18  } ;
19
20  // Constructor.
21  time24::time24( int h, int m, int s ) :
22          hours( h ), minutes( m ), seconds( s )
23  {}
24
25  // Mutator function.
26  void time24::set_time( int h, int m, int s )
27  {
28      hours = h ; minutes = m ; seconds = s ;
29  }
30
31  // Inspector function.
32  void time24::get_time( int& h, int& m, int& s ) const
33  {
34      h = hours ; m = minutes ; s = seconds ;
35  }
36
37  // Overloaded + operator.
38  time24 time24::operator+( int secs ) const
39  {
40      // Add secs to class member seconds and calculate the new time.
41      time24 temp ;
42      temp.seconds = seconds + secs ;
43      temp.minutes = minutes + temp.seconds / 60 ;
44      temp.seconds %= 60 ;
45      temp.hours = hours + temp.minutes / 60 ;
46      temp.minutes %= 60 ;
47      temp.hours %= 24 ;
48      return temp ; // Return the new time.
49  }
50
51  time24 time24::operator+( const time24& t ) const
52  {
53      // Add total seconds in t to seconds and calculate the new time.
54      time24 temp ;
55      int secs = t.hours * 3600 + t.minutes * 60 + t.seconds ;
56      temp.seconds = seconds + secs ;
57      temp.minutes = minutes + temp.seconds / 60 ;
```

```
58    temp.seconds %= 60 ;
59    temp.hours = hours + temp.minutes / 60 ;
60    temp.minutes %= 60 ;
61    temp.hours %= 24 ;
62    return temp ; // Return the new time.
63 }
64
65 int main( void )
66 {
67    int h, m, s ;
68    time24 start_time( 23, 0, 0 ) ;
69    time24 elapsed_time( 1, 2, 3 ) ;
70    time24 finish_time ;
71
72    finish_time = start_time + elapsed_time ;
73    finish_time.get_time( h, m, s ) ;
74    cout << "Finish Time is "
75        << h << ":" << m << ":" << s << endl ;
76    return 0 ;
77 }
```

The output from this program is:

```
Finish Time is 0:2:3
```

There are now two member functions called `operator+`. The second `operator+` member function is tested on line 72. This line is equivalent to:

```
finish_time = start_time.operator+( elapsed_time ) ;
```

This means that the member function `operator+()` is called for the object `start_time` using `elapsed_time` as an argument. Since the argument is a `time24` object, the `operator+()` function that takes a `time24` rather than an `int` argument is called, i.e. the `operator+()` function on lines 51 to 63.

Note that the parameter on line 51 is a reference to a `const time24` object. By using a reference, the address of the object, rather than a copy of the entire object, is passed, and using `const` guarantees that the passed object (`elapsed_time`) is not changed in the function. The `const` at the end of line 51 guarantees that the calling object (`start_time`) is not changed in the function.

One last problem still exists when using the overloaded `operator+` functions. The class can handle assignments such as `t2 = t1 + 4` and `t3 = t2 + t1`, but the assignment

```
t2 = 4 + t1
```

will cause a compiler error.

To understand the problem, consider the equivalent statement to

目前程序中有两个名为operator+的成员函数。第72行调用了第二个operator+函数。

这意味着对象start_time以对象elapsed_time为实参，调用了成员函数operator+( )。既然这里用time24类对象作为实参，那么实际被调用的就不是实参为整型的函数operator+( )，而是实参为time24类对象的函数operator+( )，即第51行至第63行的函数。

注意，程序第51行中函数的形参是一个针对const time24对象的引用。通过引用，传递的是对象的地址，而非整个对象的一个副本。同时，使用const可以保证被传送的对象（elapsed_time）在函数中不会被修改，行末的const确保调用该函数的对象（start_time）在函数中不会被修改。

在使用上面的operator+重载函数时，仍存在一个问题，即类可以

```
t2 = 4 + t1 ;
```

which is

```
t2 = 4.operator+( t1 ) ; // Invalid.
```

The data type of the object making the call to `operator+()` in this statement is an `int`, not a `time24` object.

Addition is commutative and so, `t2 = 4 + t1` should be the same as `t2 = t1 + 4`. This cannot be done with a `time24` member function, because `4` is not a `time24` object. A standalone non-class function must be used. This is demonstrated in the next program.

处理像t2 = t1 + 4或t3 = t2 + t1这样的赋值语句，但是对于赋值语句t2 = 4 + t1，将产生编译错误。在这个语句中，调用operator+( )的对象的类型为int，而不是time24对象。

加法是可交换的，所以t2 = 4 + t1应该与t2 = t1 + 4相同。但类time24的成员函数无法做到这一点，因为4不是time24对象。为此，应使用一个独立的非成员函数。

Program Example P10C

```
1   // Program Example P10C
2   // Demonstration of an overloaded + operator: version 3.
3   #include <iostream>
4   using namespace std ;
5
6   class time24 // A simple 24-hour time class.
7   {
8   public:
9     time24( int h = 0, int m = 0, int s = 0 ) ;
10    void set_time( int h, int m, int s ) ;
11    void get_time( int&h, int& m, int& s ) const ;
12    time24 operator+( int secs ) const ;
13    time24 operator+( const time24& t ) const ;
14  private:
15    int hours ;   // 0 to 23
16    int minutes ; // 0 to 59
17    int seconds ; // 0 to 59
18  } ;
19
20  // Constructor.
21  time24::time24( int h, int m, int s ) :
22          hours( h ), minutes( m ), seconds( s )
23  {}
24
25  // Mutator function.
26  void time24::set_time( int h, int m, int s )
27  {
28    hours = h ; minutes = m ; seconds = s ;
29  }
30
31  // Inspector function.
32  void time24::get_time( int& h, int& m, int& s ) const
33  {
34    h = hours ; m = minutes ; s = seconds ;
35  }
36
37  // Overloaded + operators.
```

```
38  time24 time24::operator+( int secs ) const
39  {
40    // Add secs to class member seconds and calculate the new time.
41    time24 temp ;
42    temp.seconds = seconds + secs ;
43    temp.minutes = minutes + temp.seconds / 60 ;
44    temp.seconds %= 60 ;
45    temp.hours = hours + temp.minutes / 60 ;
46    temp.minutes %= 60 ;
47    temp.hours %= 24 ;
48    return temp ; // Return the new time.
49  }
50
51  time24 time24::operator+( const time24& t ) const
52  {
53    // Add total seconds in t to seconds and calculate the new time.
54    time24 temp ;
55    int secs = t.hours * 3600 + t.minutes * 60 + t.seconds ;
56    temp.seconds = seconds + secs ;
57    temp.minutes = minutes + temp.seconds / 60 ;
58    temp.seconds %= 60 ;
59    temp.hours = hours + temp.minutes / 60 ;
60    temp.minutes %= 60 ;
61    temp.hours %= 24 ;
62    return temp ; // Return the new time.
63  }
64
65  // Non-member overloaded + operator.
66  time24 operator+( int secs, const time24& t )
67  {
68    // Add secs to t to calculate the new time.
69    time24 temp ;
70    temp = t + secs ;  // Uses the member function operator+( int ).
71    return temp ;  // Return the new time.
72  }
73
74  int main( void )
75  {
76    int h, m, s ;
77    time24 t1( 23, 59, 57 ) ;
78    time24 t2 ;
79
80    t2 = 4 + t1 ;
81    t2.get_time ( h, m, s ) ;
82    cout << "Time t2 is " << h << ":" << m << ":" << s << endl ;
83    return 0 ;
84  }
```

The output from this program is:

```
Time t2 is 0:0:1
```

The program contains a non-class standalone function called `operator+()` on lines 66 to 72. This function is not a member of the `time24` class. Like any other function it can be called in the conventional way with a statement such as:

程序的第66行至第72行包含一个独立的非成员函数operator+( )，它不是time24类的成员，因此和调用其他普通函数一样，可以用传统的方式进行调用。

```
t2 = operator+( 4, t1 ) ;
```

which is equivalent to the statement on line 80.

10.3 Rules of operator overloading
（运算符重载的规则）

All the C++ operators in appendix B with the exception of the following five operators can be overloaded:

除了下面列举的5个运算符，附录B中的所有C++运算符都可以被重载。

```
.    .*    ::    ?:    sizeof
```

The following rules apply to operator overloading:

1. New operators cannot be invented. For example, it is not possible to overload `< >` to mean 'not equal to' or `**` to mean 'to the power of'.
2. The overloaded operator must have the same number of operands as the corresponding predefined operator. For example, an overloaded equivalent operator `==` must have two operands and an overloaded `++` operator must have one operand.
3. The priority of an overloaded operator remains the same as its corresponding predefined operator. For example, regardless of whether `*` is overloaded or not, it will always have a higher precedence than `+` and `-`.
4. Overloaded operators cannot have default arguments.
5. The operators for the built-in data types, e.g. `int`, `float` and so on, cannot be redefined.

1. 不能发明新的运算符。例如，不可以重载< >来表示"不等于"或者重载**来表示"……的幂"。
2. 运算符重载后的操作数个数与原运算符的操作数个数必须相等。例如，重载运算符==必须有两个操作数，重载运算符++应只有一个操作数。
3. 运算符重载后仍保持其原有的优先级。例如，无论是否对*进行重载，它的优先级都将高于+ 和-。
4. 重载的运算符不能使用默认实参。
5. 用于内置数据类型（如int，float，等等）时，运算符的含义不能被重新定义。

10.4 Overloading ++（重载运算符 ++）

The next program includes a class member function to overload the increment operator `++`.

Program Example P10D

```
1  // Program Example P10D
2  // Program to demonstrate overloading the prefix ++ operator.
3  #include <iostream>
4  using namespace std ;
5
6  class time24 // A simple 24-hour time class.
7  {
8  public:
9    time24( int h = 0, int m = 0, int s = 0 ) ;
```

```
10    void set_time( int  h, int  m, int  s ) ;
11    void get_time( int& h, int& m, int& s ) const ;
12    time24 operator+( int secs ) const ;
13    time24 operator+( const time24& t ) const ;
14    time24 operator++() ;
15 private:
16    int hours ;    // 0 to 23
17    int minutes ; // 0 to 59
18    int seconds ; // 0 to 59
19 } ;
20
21 // Constructor.
22 time24::time24( int h, int m, int s ) :
23          hours( h ), minutes( m ), seconds( s )
24 {}
25
26 // Mutator function.
27 void time24::set_time( int  h, int m, int  s )
28 {
29    hours = h ; minutes = m ; seconds = s ;
30 }
31
32 // Inspector function.
33 void time24::get_time( int& h, int& m, int& s ) const
34 {
35    h = hours ; m = minutes ;  s = seconds ;
36 }
37
38 // Overloaded + operators.
39 time24 time24::operator+( int secs ) const
40 {
41    // Add secs to class member seconds and calculate the new time.
42    time24 temp ;
43    temp.seconds = seconds + secs ;
44    temp.minutes = minutes + temp.seconds / 60 ;
45    temp.seconds %= 60 ;
46    temp.hours = hours + temp.minutes / 60 ;
47    temp.minutes %= 60 ;
48    temp.hours %= 24 ;
49    return temp ; // Return the new time.
50 }
51
52 time24 time24::operator+( const time24& t ) const
53 {
54    // Add total seconds in t to seconds and calculate the new time.
55    time24 temp ;
56    int secs = t.hours * 3600 + t.minutes * 60 + t.seconds ;
57    temp.seconds = seconds + secs ;
58    temp.minutes = minutes + temp.seconds / 60 ;
59    temp.seconds %= 60 ;
```

```
60    temp.hours = hours + temp.minutes / 60 ;
61    temp.minutes %= 60 ;
62    temp.hours %= 24 ;
63    return temp ; // Return the new time.
64  }
65
66  // Overloaded prefix ++ operator.
67  time24 time24::operator++()
68  {
69    // Add 1 second to the calling object's seconds.
70    *this = *this + 1 ;  // Uses member function operator+( int ).
71    return *this ;  // Return updated time of calling object.
72  }
73
74  // Non-member overloaded + operator.
75  time24 operator+( int secs, const time24& t )
76  {
77    // Add secs to t to calculate the new time.
78    time24 temp ;
79    temp = t + secs ;  // Uses member function operator+( int ).
80    return temp ;  // Return the new time.
81  }
82
83  int main( void )
84  {
85    int h, m, s ;
86    time24 t1( 23, 59, 57 ) ;
87
88    ++t1 ;
89    t1.get_time ( h, m, s ) ;
90    cout << "Time t1 is " << h << ":" << m << ":" << s << endl ;
91    return 0 ;
92  }
```

The output from this program is:

```
Time t1 is 23:59:58
```

The prototype for the overloaded ++ operator is on line 14. No parameters are used and the function returns a time24 object.

The function definition for the ++ operator is on lines 67 to 72. The function makes use of the special built-in pointer this. The this pointer is available in all member functions of a class and is a pointer to the object that called the member function.

When operator++() is called on line 88, this contains the address of t1. Line 70 therefore adds 1 to t1. The member function operator+() on lines 39 to 50 (with the argument secs equal to 1) is called to do the addition.

程序的第67行至第72行是运算符++的函数定义。该函数使用了专门的内置指针this。在类的所有成员函数中都可以使用this指针，this指针指向调用成员函数的对象。

当程序的第88行调用operator++()时，this指针中包含的是对象t1的地址。因此，第70行就是将t1加1，这个加法操作是通过调用第39行至第50行的成员函数operator+()（实参secs值为1）来实现的。

10.4.1 Overloading prefix and postfix forms of ++

The increment operator ++ needs to be overloaded in both its prefix and postfix forms. For example, if t1 and t2 are time24 objects then the following two statements will produce different results

对于自增运算符++，需要对其前缀和后缀形式分别进行重载。

```
t2 = ++t1 ; // Use of prefix ++.
t2 = t1++ ; // Use of postfix ++.
```

The first statement increments t1 before assigning its value to t2. The second statement assigns t1 to t2 and then increments t1.

To distinguish between the two operator ++ functions, the postfix version uses a 'dummy' integer parameter.

The next program overloads both forms of the ++ operator.

为了区分这两种operator++函数，在其后缀形式的重载函数中使用了一个"哑的"（即虚拟的）整型形参。

Program Example P10E

```
1    // Program Example P10E
2    // Program to demonstrate overloading prefix and postfix ++.
3    #include <iostream>
4    using namespace std ;
5
6    class time24 // A simple 24-hour time class.
7    {
8    public:
9      time24( int h = 0, int m = 0, int s = 0 ) ;
10     void set_time( int  h, int  m, int  s ) ;
11     void get_time( int& h, int& m, int& s ) const ;
12     time24 operator+( int secs ) const ;
13     time24 operator+( const time24& t ) const ;
14     time24 operator++() ;       // prefix.
15     time24 operator++( int ) ; // postfix.
16   private:
17     int hours ;    // 0 to 23
18     int minutes ; // 0 to 59
19     int seconds ; // 0 to 59
20   } ;
21
22   // Constructor.
23   time24::time24( int h, int m, int s ) :
24           hours( h ), minutes( m ), seconds( s )
25   {}
26
27   // Mutator function.
28   void time24::set_time( int h, int m, int s )
29   {
30     hours = h ; minutes = m ; seconds = s ;
31   }
32
33   // Inspector function.
34   void time24::get_time( int& h, int& m, int& s ) const
```

```
35  {
36      h = hours ; m = minutes ; s = seconds ;
37  }
38
39  // Overloaded + operators.
40  time24 time24::operator+( int secs ) const
41  {
42      // Add secs to class member seconds and calculate the new time.
43      time24 temp ;
44      temp.seconds = seconds + secs ;
45      temp.minutes = minutes + temp.seconds / 60 ;
46      temp.seconds %= 60 ;
47      temp.hours = hours + temp.minutes / 60 ;
48      temp.minutes %= 60 ;
49      temp.hours %= 24 ;
50      return temp ; // Return the new time.
51  }
52
53  time24 time24::operator+( const time24& t ) const
54  {
55      // Add total seconds in t to seconds and calculate the new time.
56      time24 temp ;
57      int secs = t.hours * 3600 + t.minutes * 60 + t.seconds ;
58      temp.seconds = seconds + secs ;
59      temp.minutes = minutes + temp.seconds / 60 ;
60      temp.seconds %= 60 ;
61      temp.hours = hours + temp.minutes / 60 ;
62      temp.minutes %= 60 ;
63      temp.hours %= 24 ;
64      return temp ; // Return the new time.
65  }
66
67  // Overloaded prefix ++ operator.
68  time24 time24::operator++()
69  {
70      // Add 1 second to the calling object's seconds.
71      *this = *this + 1 ; // Uses member function operator+( int ).
72      return *this ; // Return updated time of calling object.
73  }
74
75  // Overloaded postfix ++ operator.
76  time24 time24::operator++( int ) ←——— Dummy integer parameter.
77  {
78      // Save calling object before incrementing seconds.
79      time24 temp ;
80      temp = *this ;
81      *this = *this + 1 ;  // Uses operator+( int ),
82                           // could also use ++(*this).
83      return temp ;  // Return the saved calling object.
84  }
85
```

```
86   // Non-member overloaded + operator.
87   time24 operator+( int secs, const time24& t )
88   {
89     // Add secs to t to calculate the new time.
90     time24 temp ;
91     temp = t + secs ;  // Uses member function operator+( int ).
92     return temp ;  // Return the new time.
93   }
94
95   int main( void )
96   {
97     int h, m, s ;
98     time24 t1( 23, 59, 57 ) ;
99     time24 t2 ;
100
101    t2 = t1++ ; // Test postfix ++
102    t1.get_time ( h, m, s ) ;
103    cout << "Using postfix ++: " << "time t1 is: "
104        << h << ":" << m << ":" << s ;
105    t2.get_time( h, m, s ) ;
106    cout << ", time t2 is: "
107        << h << ":" << m << ":" << s << endl ;
108
109    t1.set_time (23, 59, 57 ) ;  // Reset the time.
110
111    t2 = ++t1 ;  // Test prefix ++
112    t1.get_time ( h, m, s ) ;
113    cout << "Using prefix  ++: " << "time t1 is: "
114        << h << ":" << m << ":" << s ;
115    t2.get_time ( h, m, s ) ;
116    cout << ", time t2 is: "
117        << h << ":" << m << ":" << s << endl ;
118    return 0 ;
119  }
```

The output from this program is:

```
Using postfix ++: time t1 is: 23:59:58, time t2 is: 23:59:57
Using prefix  ++: time t1 is: 23:59:58, time t2 is: 23:59:58
```

Both versions of the ++ operator are tested in main().

Lines 101 to 107 test and display the result of using the postfix operator ++ and lines 111 to 117 test and display the result of using the prefix operator ++. The value of t1 is identical in both tests, but t2 is assigned two different values.

The postfix version of operator ++ is defined in lines 76 to 84. The original value of the time24 object that called the member function is stored in temp. Only after saving the original value of the time24 object is its value incremented. The original value of the time24 object (stored in temp) is then returned.

第101行至第107行测试并显示了使用后缀运算符++的结果，第111行至第117行测试并显示了使用前缀运算符++的结果。在这两组测试中，t1的值是相同的，但是t2被赋予了不同的值。

第76行至第84行定义了后缀运算符++的重载函数。调用该成员函数的time24对象的初值存储于temp中。仅在保存time24对象的初值以后，才将其值加1，最后返

10.4.2 Improving the prefix ++ operator member function

The `return` statement on line 72 of program P10E returns a copy of the object. It would be more efficient (especially for large objects) to return a reference to the object instead. To do this requires only two small changes to the class. The operator++ function header on line 68 changes to:

```
time24& time24::operator++()
```

and the function prototype on line 14 changes to:

```
time24& operator++() ;
```

The same improvement cannot be made for the overloaded postfix version of ++. The reason for this is that the object returned on line 83 is the local object `temp`, which will be out of scope and therefore undefined when the function ends. Returning a reference to an undefined object will cause problems, so a copy of `temp` has to be returned.

The overloaded + operator on lines 40 to 51 and the non-member overloaded + operator on line 87 to 93 also return a copy of a local object `temp`. Attempting to return a reference to `temp` will cause problems.

10.5 Overloading relational operators
（重载关系运算符）

An equality operator (==) for the `time24` class can be defined to test two `time24` objects for equality.

The next program uses an overloaded == operator to verify that `t1` and `t2` are equal after the statement

```
t1 = ++t2 ;
```

Program Example P10F

```
1    // Program Example P10F
2    // Program to demonstrate the overloading of the == operator.
3    #include <iostream>
4    using namespace std ;
5
6    class time24  // A simple 24-hour time class.
7    {
8    public:
9      time24( int h = 0, int m = 0, int s = 0 ) ;
10     void set_time( int h, int m, int s ) ;
11     void get_time( int&h, int& m, int& s ) const ;
12     time24 operator+( int secs ) const ;
13     time24 operator+( const time24& t ) const ;
```

回存于 temp 中的 time24 对象的初值。

程序 P10E 的第 72 行的 return 语句返回了对象的副本。如果返回的是该对象的引用，则效率会更高（尤其对于大的对象）。只需对该类做两个小的改动即可实现。

但是对于重载后缀运算符++的成员函数，不能做这样的改写，因为第 83 行返回的对象是一个局部对象 temp，当其所在函数调用结束后，由于超出了该局部对象的作用域而成为未定义的对象。返回一个未定义的对象的引用将产生错误，因此只能返回 temp 对象的副本。

第 40 行至第 51 行的重载运算符+的成员函数及第 87 行至第 93 行重载运算符+的非成员函数也是返回局部对象 temp 的副本。试图返回该对象的引用是错误的。

```
14    time24& operator++() ;        // prefix.
15    time24 operator++( int ) ;   // postfix.
16    bool operator == ( const time24& t ) const ;
17  private:
18    int hours ;    // 0 to 23
19    int minutes ; // 0 to 59
20    int seconds ; // 0 to 59
21  } ;
22
23  // Constructor.
24  time24::time24( int h, int m, int s ) :
25          hours( h ), minutes( m ), seconds( s )
26  {}
27
28  // Mutator function.
29  void time24::set_time( int h, int m, int s )
30  {
31    hours = h ; minutes = m ; seconds = s ;
32  }
33
34  // Inspector function.
35  void time24::get_time( int& h, int& m, int& s ) const
36  {
37    h = hours ; m = minutes ; s = seconds ;
38  }
39
40  // Overloaded + operators.
41  time24 time24::operator+( int secs ) const
42  {
43    // Add secs to class member seconds and calculate the new time.
44    time24 temp ;
45    temp.seconds = seconds + secs ;
46    temp.minutes = minutes + temp.seconds / 60 ;
47    temp.seconds %= 60 ;
48    temp.hours = hours + temp.minutes / 60 ;
49    temp.minutes %= 60 ;
50    temp.hours %= 24 ;
51    return temp ;   // Return the new time.
52  }
53
54  time24 time24::operator+( const time24& t ) const
55  {
56    // Add total seconds in t to seconds and calculate the new time.
57    time24 temp ;
58    int secs = t.hours * 3600 + t.minutes * 60 + t.seconds ;
59    temp.seconds = seconds + secs ;
60    temp.minutes = minutes + temp.seconds / 60 ;
61    temp.seconds %= 60 ;
62    temp.hours = hours + temp.minutes / 60 ;
63    temp.minutes %= 60 ;
```

```
64    temp.hours %= 24 ;
65    return temp ;  // Return the new time.
66  }
67
68  // Overloaded prefix ++ operator
69  time24& time24::operator++()
70  {
71    // Add 1 second to the calling object's seconds.
72    *this = *this + 1 ; // Uses member function operator+( int ).
73    return *this ; // Return updated time of calling object.
74  }
75
76  // Overloaded postfix ++ operator.
77  time24 time24::operator++( int )
78  {
79    // Save calling object before incrementing seconds.
80    time24 temp ;
81    temp = *this ;
82    *this = *this + 1 ;  // Uses operator+( int ),
83                         // could also use ++(*this).
84    return temp ;  // Return the saved calling object.
85  }
86
87  // Overloaded equality operator ==.
88  bool time24::operator == ( const time24& t ) const
89  {
90    if ( hours == t.hours &&
91         minutes == t.minutes &&
92         seconds == t.seconds )
93      return true ;
94    else
95      return false ;
96  }
97
98  // Non-member overloaded + operator.
99  time24 operator+( int secs, const time24& t )
100 {
101   // Add secs to t to calculate the new time.
102   time24 temp ;
103   temp = t + secs ; // Uses member function operator+( int ).
104   return temp ; // Return the new time.
105 }
106
107 int main( void )
108 {
109   int h, m, s ;
110   time24 t1( 23, 59, 57 ) ;
111   time24 t2 ;
112
113   t2 = ++t1 ;  // t1 and t2 should be equal.
```

```
114    t1.get_time ( h, m, s ) ;
115    cout << "t1 is "
116         << h << ":" << m << ":" << s ;
117    t2.get_time ( h, m, s ) ;
118    cout << ", t2 is "
119         << h << ":" << m << ":" << s << endl ;
120
121    if ( t1 == t2 ) // Test equality operator ==.
122      cout << "t1 and t2 are equal. Prefix ++ is working."
123           << endl ;
124    else
125      cout << "t1 and t2 are not equal. Prefix ++ is not working."
126           << endl ;
127    return 0 ;
128 }
```

The output from this program is:

```
t1 is 23:59:58, t2 is 23:59:58
t1 and t2 are equal. Prefix ++ is working.
```

The overloaded operator == for the time24 class on lines 88 to 96 returns a bool value of true if the data members of the calling object (t1) and the argument (t2) are equal, otherwise false is returned.

10.6   Overloading << and >>（重载运算符 << 和 >>）

Just like any other operator, the insertion and extraction operators << and >> can be overloaded. This enables the time24 class to have its own insertion and extraction operators << and >>.

与其他运算符一样，流插入运算符<<和流提取运算符>>也能被重载。

Program Example P10G

```
1   // Program Example P10G
2   // Program to demonstrate the overloading of << and >>.
3   #include <iostream>
4   #include <iomanip>
5   using namespace std ;
6
7   class time24  // A simple 24-hour time class.
8   {
9   public:
10    time24( int h = 0, int m = 0, int s = 0 ) ;
11    void set_time( int  h, int  m, int  s ) ;
12    void get_time( int& h, int& m, int& s ) const ;
13    time24 operator+( int secs ) const ;
14    time24 operator+( const time24& t ) const ;
15    time24& operator++() ;       // prefix.
16    time24 operator++( int ) ;   // postfix.
17    bool operator == ( const time24& t ) const ;
18  private:
```

```
19    int hours ;    // 0 to 23
20    int minutes ; // 0 to 59
21    int seconds ; // 0 to 59
22  } ;
23
24  // Constructor.
25  time24::time24( int h, int m, int s ) :
26          hours( h ), minutes( m ), seconds( s )
27  {}
28
29  // Mutator function.
30  void time24::set_time( int  h, int m, int  s )
31  {
32    hours = h ; minutes = m ; seconds = s ;
33  }
34
35  // Inspector function.
36  void time24::get_time( int& h, int& m, int& s ) const
37  {
38    h = hours ; m = minutes ;  s = seconds ;
39  }
40
41  // Overloaded + operators.
42  time24 time24::operator+( int secs ) const
43  {
44    // Add secs to class member seconds and calculate the new time.
45    time24 temp ;
46    temp.seconds = seconds + secs ;
47    temp.minutes = minutes + temp.seconds / 60 ;
48    temp.seconds %= 60 ;
49    temp.hours = hours + temp.minutes / 60 ;
50    temp.minutes %= 60 ;
51    temp.hours %= 24 ;
52    return temp ;  // Return the new time.
53  }
54
55  time24 time24::operator+( const time24& t ) const
56  {
57    // Add total seconds in t to seconds and calculate the new time.
58    time24 temp ;
59    int secs = t.hours * 3600 + t.minutes * 60 + t.seconds ;
60    temp.seconds = seconds + secs ;
61    temp.minutes = minutes + temp.seconds / 60 ;
62    temp.seconds %= 60 ;
63    temp.hours = hours + temp.minutes / 60 ;
64    temp.minutes %= 60 ;
65    temp.hours %= 24 ;
66    return temp ;  // Return the new time.
67  }
68
69  // Overloaded prefix ++ operator.
```

```
70   time24& time24::operator++()
71   {
72     // Add 1 second to the calling object's seconds.
73     *this = *this + 1 ;   // Uses member function operator+( int ).
74     return *this ;   // Return updated time of calling object.
75   }
76
77   // Overloaded postfix ++ operator.
78   time24 time24::operator++( int )
79   {
80     // Save calling object before incrementing seconds.
81     time24 temp ;
82     temp = *this ;
83     *this = *this + 1 ;   // Uses operator+( int ),
84                           // could also use ++(*this).
85     return temp ;   // Return the saved calling object.
86   }
87
88   // Overloaded equality operator ==.
89   bool time24::operator == ( const time24& t ) const
90   {
91     if ( hours == t.hours &&
92           minutes == t.minutes &&
93           seconds == t.seconds )
94       return true ;
95     else
96       return false ;
97   }
98
99   // Non-member overloaded + operator.
100   time24 operator+( int secs, const time24& t )
101  {
102    // Add secs to t to calculate the new time.
103    time24 temp ;
104    temp = t + secs ;   // Uses member function operator+( int ).
105    return temp ;   // Return the new time.
106  }
107
108  // Non-member overloaded << operator.
109  ostream& operator<<( ostream& os, const time24& t )
110  {
111    // Format time24 object, precede single digits with a 0.
112    int h, m, s ;
113    t.get_time( h, m, s ) ;
114    os << setfill( '0' )
115       << setw( 2 ) << h << ":"
116       << setw( 2 ) << m << ":"
117       << setw( 2 )<< s << endl ;
118    return os ;
119  }
120
```

```
121 // Non-member overloaded >> operator.
122 istream& operator>>( istream& is, time24& t )
123 {
124     // Input a time24 object data in the format h:m:s.
125     int h, m, s ;
126     do
127     {
128         is >> h ;
129     }
130     while ( h < 0 || h > 23 ) ;
131     // Ignore the separator.
132     is.ignore( 1 ) ;
133     do
134     {
135         is >> m ;
136     }
137     while ( m < 0 || m > 60 ) ;
138     // Ignore the separator.
139     is.ignore( 1 ) ;
140     do
141     {
142         is >> s ;
143     }
144     while ( s < 0 || s > 60 ) ;
145     t.set_time ( h, m, s ) ;
146     return is ;
147 }
148
149 int main( void )
150 {
151     time24 t1( 1, 2, 3 ) ;
152     time24 t2 ( 10, 10, 10 ) ;
153
154     cout << t1 << t2 ;
155
156     time24 t3 ;
157     cin  >> t3 ;
158     cout << t3 ;
159     return 0 ;
160 }
```

A sample run of this program is:

```
01:02:03
10:10:10
4:5:6        ◄───────── Input a time.
04:05:06
```

Both >> and << are implemented as non-member functions. The headers for these functions on lines 109 and 122 look complicated and require an explanation.

Firstly, cout is an object of the class ostream and cin is an object of the class istream. These objects and classes are defined in the header file iostream included in the program on line 3.

A statement such as

```
cout << "hello" ;
```

results in a change in the output stream object cout. Any function that uses an output stream will modify it, which means that the output stream must be passed by reference, rather than by value, to a function. This is why the first parameter on line 109 is a reference to an object of type ostream.

The second parameter t is declared as a reference to a const time24 object for the purpose of efficiency. By using a reference, the address of the object, rather than a copy of the object is passed to t. Using a const ensures that the time24 object that t refers to cannot be modified within the function.

The return type on line 109 is a reference to the output stream which is returned on line 118. To understand the reason for this, consider a statement such as:

```
cout << t1 << t2 ;
```

The first part of this statement (cout << t1) displays the value of t1 and also returns a reference to cout enabling the next part of the statement (<< t2) to also use cout.

If the multiple uses of << in the same statement were not required, the return type on line 109 could be simply void and line 118 could be omitted.

There is a similar explanation for the function header for >> on line 122. The first parameter must be a reference to istream to enable >> to be used more than once in the same statement. The second parameter t must be a reference, since the object that t refers to will be modified by the function.

The function itself inputs values for the private data members of a time24 class object, performing some elementary validation on the input values.

cout是类ostream的对象，cin是类istream的对象。这些对象和类是在程序第3行包含的头文件iostream中定义的。

任何使用输出流的函数都会改变cout，因此向函数传递的必须是输出流的引用，而不能传值。这也是第109行中第一个参数为ostream对象的引用的原因。

为了提高效率，将第二个参数t也声明为const time24对象的引用。通过使用引用，将对象的地址而并非对象的副本传递给t。使用const是为了保证t所指向的time24对象在函数中不会被修改。

第109行函数的返回类型是一个由第118行返回的输出流对象的引用。为了分析返回引用的原因，考虑语句cout << t1 << t2;的第一部分（cout << t1）输出t1的值，并返回一个cout的引用，使得语句的后面部分（<< t2）可以继续使用cout。

如果不要求在同一条语句中多次使用运算符<<，那么可以把第109行的函数返回类型改为void，同时删除第118行的语句。同理，第122行重载运算符>>的函数头部中的第一个参数必须是istream的引用，使>>可以在同一条语句中被多次使用。第二个参数t也必须为一个引用，因为t所指向的对象要在函数中被修改。

10.7　Conversion operators（转换运算符）

A conversion operator function is used to convert from a class object to a built-in data type or to another class object. The conversion operator member function has the same name as the data type to which the class is to be converted.

Conversion operators differ from other overloaded operators in two ways.

Firstly, a conversion operator takes no arguments.

Secondly, a conversion operator has no return value, not even `void`. The return value from a conversion operator is inferred from the operator's name. For example, to convert a `time24` object to an `int` a conversion operator named `operator int` is defined in the class. This is demonstrated in the next program.

转换运算符函数用于将一个类对象转换成内置数据类型或其他类对象。转换运算符成员函数与它要转换成的数据类型具有相同的名字。

转换运算符函数与其他重载运算符函数有两点不同：

第一，转换运算符函数无须实参。

第二，转换运算符函数没有返回类型，甚至void也不能使用。可以根据转换运算符函数的名字推出函数的返回类型。例如，如果要将一个time24对象转换为int类型，则在类中定义一个名为operator int的转换运算符。

Program Example P10H

```
1    // Program Example P10H
2    // Demonstration of a class conversion operator.
3    #include <iostream>
4    #include <iomanip>
5    using namespace std ;
6
7    class time24 // A simple 24 hour time class.
8    {
9    public:
10     time24( int h = 0, int m = 0, int s = 0 ) ;
11     void set_time( int  h, int  m, int  s ) ;
12     void get_time( int& h, int& m, int& s ) const ;
13     time24 operator+( int secs ) const ;
14     time24 operator+( const time24& t ) const ;
15     time24& operator++() ;        // prefix.
16     time24 operator++( int ) ;  // postfix.
17     bool operator == ( const time24& t ) const ;
18     operator int() ;
19    private:
20     int hours ;   // 0 to 23
21     int minutes ; // 0 to 59
22     int seconds ; // 0 to 59
23    } ;
24
25    // Constructor.
26    time24::time24( int h , int m, int s ) :
27            hours( h ), minutes( m ), seconds( s )
28    {}
29
30    // Mutator function.
31    void time24::set_time( int  h, int m, int  s )
32    {
```

```
33     hours = h ; minutes = m ; seconds = s ;
34   }
35
36   // Inspector function.
37   void time24::get_time( int& h, int& m, int& s ) const
38   {
39     h = hours ; m = minutes ; s = seconds ;
40   }
41
42   // Overloaded + operators.
43   time24 time24::operator+( int secs ) const
44   {
45     // Add secs to class member seconds and calculate new time.
46     time24 temp ;
47     temp.seconds = seconds + secs ;
48     temp.minutes = minutes + temp.seconds / 60 ;
49     temp.seconds %= 60 ;
50     temp.hours = hours + temp.minutes / 60 ;
51     temp.minutes %= 60 ;
52     temp.hours %= 24 ;
53     return temp ; // Return the new time.
54   }
55
56   time24 time24::operator+( const time24& t ) const
57   {
58     // Add total seconds in t to seconds and calculate new time.
59     time24 temp ;
60     int secs = t.hours * 3600 + t.minutes * 60 + t.seconds ;
61     temp.seconds = seconds + secs ;
62     temp.minutes = minutes + temp.seconds / 60 ;
63     temp.seconds %= 60 ;
64     temp.hours = hours + temp.minutes / 60 ;
65     temp.minutes %= 60 ;
66     temp.hours %= 24 ;
67     return temp ; // Return the new time.
68   }
69
70   // Overloaded prefix ++ operator
71   time24& time24::operator++()
72   {
73     // add 1 second to the calling object's seconds.
74     *this = *this + 1 ; // Uses member function operator+( int ).
75     return *this ; // Return updated time of calling object.
76   }
77
78   // Overloaded postfix ++ operator.
79   time24 time24::operator++( int )
80   {
81     // Save calling object before incrementing seconds.
82     time24 temp ;
```

```
83    temp = *this ;
84    *this = *this + 1 ;  // Uses operator+( int ),
85                         // could also use ++(*this).
86    return temp ;  // Return the saved calling object.
87  }
88
89  // Overloaded equality operator ==.
90  bool time24::operator == ( const time24& t ) const
91  {
92    if ( hours == t.hours &&
93         minutes == t.minutes &&
94         seconds == t.seconds )
95      return true ;
96    else
97      return false ;
98  }
99
100 // Conversion operator from time24 to an integer.
101 time24::operator int()
102 {
103   int no_of_seconds = hours * 3600 + minutes * 60 + seconds ;
104   return no_of_seconds ;
105 }
106
107 // Non-member overloaded + operator.
108 time24 operator+( int secs, const time24& t )
109 {
110   // Add secs to t to calculate the new time.
111   time24 temp ;
112   temp = t + secs ; // Uses member function operator+( int ).
113   return temp ; // Return the new time.
114 }
115
116 // Non-member overloaded << operator.
117 ostream& operator<<( ostream& os, const time24& t )
118 {
119   // Format time24 object, precede single digits with a 0.
120   int h, m, s ;
121   t.get_time( h, m, s ) ;
122   os << setfill( '0' )
123      << setw( 2 ) << h << ":"
124      << setw( 2 ) << m << ":"
125      << setw( 2 )<< s << endl ;
126   return os ;
127 }
128
129 // Non-member overloaded >> operator.
130 istream& operator>>( istream& is, time24& t )
131 {
132   // Input a time24 object data in the format h:m:s.
```

```
133    int h, m, s ;
134    do
135      is >> h ;
136    while ( h < 0 || h > 23 ) ;
137    // Ignore the separator.
138    is.ignore( 1 ) ;
139    do
140      is >> m ;
141    while ( m < 0 || m > 60 ) ;
142    // Ignore the separator.
143    is.ignore( 1 ) ;
144    do
145      is >> s ;
146    while ( s < 0 || s > 60 ) ;
147    t.set_time ( h, m ,s ) ;
148    return is ;
149 }
150
151 int main( void )
152 {
153    time24 t( 1, 2, 3 ) ;
154    int s ;
155
156    s = t ; // Conversion from a time24 data type to an int data type.
157    cout << "Time = " << t
158        << "Equivalent number of seconds = " << s << endl ;
159    return 0 ;
160 }
```

The output from this program is:

```
Time = 01:02:03
Equivalent number of seconds = 3723
```

The `time24` object t is converted to an `int` value on line 156. This statement is equivalent to the statement `s = int( t )` and results in a call to the conversion `operator()` on lines 101 to 105. This function must be a class member function that has no return type, even though it returns a value. The return type is implied by the name of the conversion operator.

第156行将time24对象t转换成一个int类型的值。此语句等价于s = int(t)，引起第101行至第105行的转换运算符函数operator( )被调用。即使它返回一个值，该函数仍必须声明为一个无返回类型的类成员函数。转换运算符函数的名字就隐式地说明了其返回类型。

10.8 Use of `friend` functions（使用友元函数）

The values of the `private` members `hours`, `minutes` and `seconds` of the `time24` class are accessed using the class member function `get_time()` and given values using the member function `set_time()`. To access the private data members of a class directly in a non-member function, the function must be declared to be a `friend` of the class.

To declare the non-member functions operator >> and operator << as

time24类的私有成员hours、minutes和seconds的值可以使用类成员函数get_time( )来读取，使用set_time( )来赋值。在一个非成员函数中，要想直接访问类的私有数据成员，该函数必须声明为这个类的一个友元。

为了将非成员函数operator >>和

friends of the `time24` class, the following declarations are inserted in the class:

```
friend ostream& operator<<( ostream& os, const time24& t ) ;
friend istream& operator>>( istream& is, time24& t ) ;
```

These declarations can be placed anywhere in the class, but they are usually placed in the `public` section.

The overhead involved in calling the member functions `get_time()` and `set_time()` can now be eliminated by direct reference to the `private` data members `hours`, `minutes` and `seconds`. The function `operator<<`, for example, can now be re-written as:

```
ostream& operator<<( ostream& os, const time24& t )
{
  // Format time24 object, precede single digits with a 0.
  os << setfill( '0' )
    << setw( 2 ) << t.hours << ":"
    << setw( 2 ) << t.minutes << ":"
    << setw( 2 ) << t.seconds << endl ;
  return os ;
}
```

Friend functions override a basic principle of object-oriented programming - that of data hiding. Friends of a class have access to all the private data of a class and their use should be minimised where possible.

operator <<声明为time24类的友元，在类中应加入下列声明：

这些声明可以放在类中的任何地方，但通常将其放在public部分。

通过直接引用私有数据成员hours、minutes和seconds，可以将前面程序中对成员函数get_time( )和set_time( )的调用删除。

友元函数破坏了面向对象程序的基本准则，即数据隐藏。类的友元可以访问类的所有私有数据，应尽量减少友元的使用。

10.9　Overloading the assignment operator =
（重载赋值运算符=）

10.9.1　A class with a pointer data member

Consider a class for recording the transactions made by a bank customer. To simplify matters, the class `private` data members are the bank account number, the number of transactions made and the amount of each transaction. Each bank account will have a different number of transactions, but the maximum number of transactions is set to one hundred, say.

A first attempt at writing the class might be like this:

请设计一个类来记录银行客户的交易。为了简化问题，类的私有数据成员只有三个：银行账号、交易的次数和每笔交易额。每个银行账号都有一个不同的交易次数，交易的最大次数均设置为100。

```
class transactions
{
public:
  transactions () ; // Constructor.
  ...
private:
  int account_number ;
  int number_of_transactions ;
  float transaction_amounts[100] ;
} ;
```

Each `transactions` object has storage for a hundred floating-point transaction values. If there are no transactions or a hundred transactions, the array size is fixed at 100. This would be a very inefficient for a large number of customers and would restrict the number of transactions to 100. It would be better to use dynamic memory allocation to allocate the exact storage required by each account. This requires a class to have a pointer data member and is demonstrated in the next program.

每个transactions对象可以存储100个浮点型交易额。无论交易的数量是多是少，数组大小都固定为100。当客户数量很多时，这样做的效率是很低的，并且交易的次数被限制为不能超过100。更好的方法是使用动态内存分配为每个账号分配恰好满足其大小需求的存储空间。这需要一个含有指针数据成员的类，用下面程序加以说明。

Program Example P10I

```cpp
1   // Program Example P10I
2   // Demonstration of a class with a pointer data member.
3   #include <iostream>
4   using namespace std ;
5
6   class transactions  // A class containing a pointer data member.
7   {
8   public:
9     transactions( int ac_num, int num_transactions,
10                  const float transaction_values[] ) ;
11    // Purpose: Constructor.
12    ~transactions() ;
13    // Purpose: Destructor.
14    void display( void ) const ;
15    // Purpose: Display transactions.
16  private:
17    int account_number ;
18    int number_of_transactions ;
19    float *transaction_amounts ; // Pointer to array of transactions.
20  } ;
21
22  // Constructor.
23  transactions::transactions( int ac_num, int num_transactions,
24                           const float transaction_values[] )
25  {
26    account_number = ac_num ;
27    number_of_transactions = num_transactions ;
28    // Allocate storage for the transactions.
29    transaction_amounts = new float[number_of_transactions] ;
30    for ( int i = 0 ; i < number_of_transactions ; i++ )
31      transaction_amounts[i] = transaction_values[i] ;
32  }
33
34  // Destructor.
35  transactions::~transactions()
36  {
37    delete[] transaction_amounts ;
38  }
39
```

```
40 void transactions::display( void ) const
41 {
42   cout << " Account Number " << account_number << " Transactions: ";
43   for ( int i = 0 ; i < number_of_transactions ; i++ )
44     cout << transaction_amounts[i] << ' ' ;
45   cout << endl ;
46 }
47
48 int main( void )
49 {
50   // Construct a transactions object
51   // with account number 1 and 5 transactions.
52   float trans_amounts1[] = { 25.67, 6.12, 19.86, 23.41, 1.21 } ;
53   transactions trans_object1( 1, 5, trans_amounts1 ) ;
54   trans_object1.display() ;
55   return 0 ;
56 }
```

The output from this program is:

```
Account Number 1 Transactions: 25.67 6.12 19.86 23.41 1.21
```

Line 53 creates a `transactions` object with account number 1 and five transactions. The exact storage required for the transaction amounts is allocated using `new` in the constructor on line 29.

Lines 35 to 38 is a special class member function called a *destructor*. A destructor always has the same name as the class itself, but is preceded with a tilde(~).

A destructor is generally used in a class that has a pointer data member that points to memory allocated by the class. In such a case, a destructor can be used to free the memory allocated by the class. In this example, line 37 frees the memory allocated by the constructor on line 29.

Just as a class constructor is called automatically when a class object is created, a class destructor is called automatically just before the class object is destroyed. An object is destroyed, for example, when the program ends or a function in which the object is created ends.

A class can only have one destructor. A destructor has no parameters and no return type.

第 35 行至第 38 行是一个特殊的成员函数，称为析构函数。析构函数总是与其所属的类同名，只是其名字前面多了一个波浪号（~）。

如果类中含有指针数据成员，指向由该类分配的内存空间，则通常需要使用析构函数。在这种情况下，析构函数用来释放由类分配的内存。

正如创建类对象时会自动调用构造函数一样，在类对象被撤销前也会自动调用析构函数。例如，当程序结束或创建该对象的函数执行结束时，该对象被撤销。

一个类只能有一个析构函数。析构函数没有形参，也没有返回类型。

10.9.2　Assigning one object to another

When assigning one object to another object of the same type, member-wise assignment is performed by default. For example, if `t1` and `t2` are `time24` objects then the statement

当两个相同类型的对象互相赋值时，默认情形是成员逐项赋值。

```
t1 = t2 ;
```

has the effect of copying each member one by one as in:

```
t1.hours = t2.hours ;
```

```
t1.minutes = t2.minutes ;
t1.seconds = t2.seconds ;
```

Each data member of t2 is copied byte by byte to its corresponding data member in t1. The default assignment operator works well for most classes, but fails to work properly for classes that allocate memory for one or more of its data members.

对于大多数的类而言，默认的赋值运算符都能正确赋值，但是当类为它的一个或多个数据成员动态分配内存后，却会产生一些问题。

Program Example P10J

```
1  // Program Example P10J
2  // Program to demonstrate the default assignment operator problem
3  // for a class with a pointer data member.
4  #include <iostream>
5  using namespace std ;
6
7  class transactions // A class containing a pointer data member.
8  {
9  public:
10   transactions( int ac_num, int num_transactions,
11                const float transaction_values[] ) ;
12   // Purpose: Constructor.
13   ~transactions() ;
14   // Purpose: Destructor.
15   void modify_transaction( int transaction_index, float amount ) ;
16   // Purpose: Modify a transaction amount.
17   // Arguments: transaction_index - transaction to change.
18   //            amount - new transaction amount.
19   void display( void ) const ;
20   // Purpose: Display transactions.
21  private:
22   int account_number ;
23   int number_of_transactions ;
24   float *transaction_amounts ; // Pointer to array of transactions.
25  } ;
26
27  // Constructor.
28  transactions::transactions( int ac_num, int num_transactions,
29                           const float transaction_values[] )
30  {
31   account_number = ac_num ;
32   number_of_transactions = num_transactions ;
33   // Allocate storage for the transactions.
34   transaction_amounts = new float[number_of_transactions] ;
35   for ( int i = 0 ; i < number_of_transactions ; i++ )
36     transaction_amounts[i] = transaction_values[i] ;
37  }
38
39  // Destructor.
40  transactions::~transactions()
41  {
```

```
42      delete[] transaction_amounts ;
43  }
44
45  // Modify a transaction.
46  void   transactions::modify_transaction( int transaction_index,
47                                   float amount )
48  {
49    // Modify the transaction amount if the array index is valid
50    if ( transaction_index >= 0 &&
51          transaction_index < number_of_transactions )
52      transaction_amounts[transaction_index] = amount ;
53  }
54
55  void transactions::display( void ) const
56  {
57    cout << " Account Number " << account_number << " Transactions: ";
58    for ( int i = 0 ; i < number_of_transactions ; i++ )
59      cout << transaction_amounts[i] << ' ' ;
60    cout << endl ;
61  }
62
63  int main( void )
64  {
65    float trans_amounts1[] = { 25.67, 6.12, 19.86, 23.41, 1.21 } ;
66    transactions trans_object1( 1, 5, trans_amounts1 ) ;
67    float trans_amounts2[] = { 16.82, 8.45, 56.19 } ;
68    transactions trans_object2( 2, 3, trans_amounts2 ) ;
69
70    cout << "Before assignment:" << endl ;
71    cout << "trans_object1:" << endl ;
72    trans_object1.display() ;
73    cout << "trans_object2:" << endl ;
74    trans_object2.display() ;
75
76    trans_object2 = trans_object1 ;
77
78    cout << endl << "After assignment:" << endl ;
79    cout << "trans_object1:" << endl ;
80    trans_object1.display() ;
81    cout << "trans_object2:" << endl ;
82    trans_object2.display() ;
83
84    // Change first transaction amount of trans_object1 to 9.99
85    trans_object1.modify_transaction( 0, 9.99 ) ;
86
87    cout << endl
88        << "After changing first transaction amount of"
89        << " trans_object1 to 9.99" << endl ;
90    cout << "trans_object1:" << endl ;
91    trans_object1.display() ;
```

```
92    cout << "trans_object2:" << endl ;
93    trans_object2.display() ;
94    return 0 ;
95  }
```

The output from this program is:

```
Before assignment:
trans_object1:
Account Number 1 Transactions: 25.67 6.12 19.86 23.41 1.21
trans_object2:
Account Number 2 Transactions: 16.82 8.45 56.19

After assignment:
trans_object1:
Account Number 1 Transactions: 25.67 6.12 19.86 23.41 1.21
trans_object2:
Account Number 1 Transactions: 25.67 6.12 19.86 23.41 1.21

After changing first transaction amount of trans_object1 to 9.99
trans_object1:
Account Number 1 Transactions: 9.99 6.12 19.86 23.41 1.21
trans_object2:
Account Number 1 Transactions: 9.99 6.12 19.86 23.41 1.21
```

After line 68, a sketch of memory might be as shown below (see Figure 10.1). Fictitious memory addresses are to the left of each memory location.

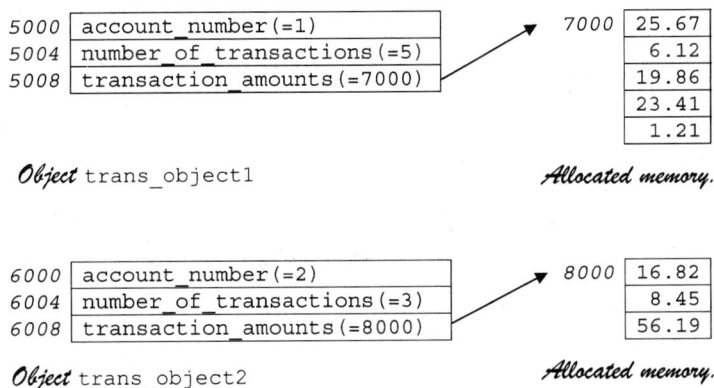

Figure 10.1

On line 76, `trans_object1` is copied to `trans_object2`. By default, a member-wise copy occurs. This means that the block of memory occupied by `trans_object1` is copied to the memory block occupied by `trans_object2`, byte by byte.

This results in the following changes in memory (see Figure 10.2):

第 76 行将 trans_object1 复制给 trans_object2。默认情况下，成员逐项复制。这意味着 trans_object1 所占用的内存块被逐个字节地复制给 trans_object2 所占用的内存块。这导致内存中发生了如下变化。

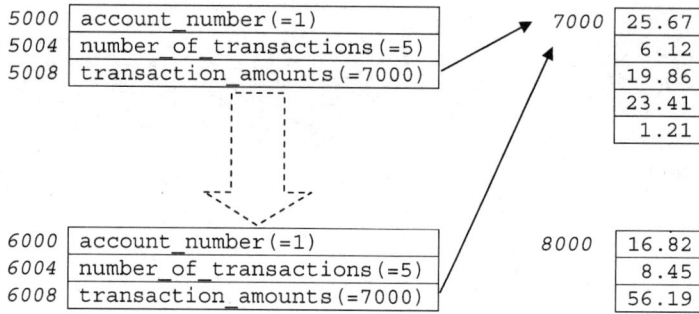

5000	account_number(=1)
5004	number_of_transactions(=5)
5008	transaction_amounts(=7000)

7000	25.67
	6.12
	19.86
	23.41
	1.21

6000	account_number(=1)
6004	number_of_transactions(=5)
6008	transaction_amounts(=7000)

8000	16.82
	8.45
	56.19

Figure 10.2

The contents of each data member of trans_object1 have been copied to the corresponding data member in trans_object2. Note that the contents of the pointer was copied and not the actual memory location it points to. This type of copy is called a *shallow copy*.

As a result of the shallow copy, the transaction_amounts pointer in the two objects are pointing to the same memory block. A *deep copy* is required, in which the memory pointed to by a pointer, rather than just the pointer itself, is copied.

Line 85 changes the first transaction amount of trans_object1 resulting in the following memory change (see Figure 10.3):

trans_object1的每个数据成员中的内容都被复制给trans_object2相应的数据成员。注意，只复制了指针的内容（即地址），而指针所指向的实际内存单元中的数据并没有被复制。这种类型的复制称为浅拷贝。

经过浅拷贝，两个对象各自的transaction_amounts指针指向了相同的内存区。当不仅要复制指针本身的值，而且还要复制该指针所指向的实际内存单元的值时，需要使用深拷贝。

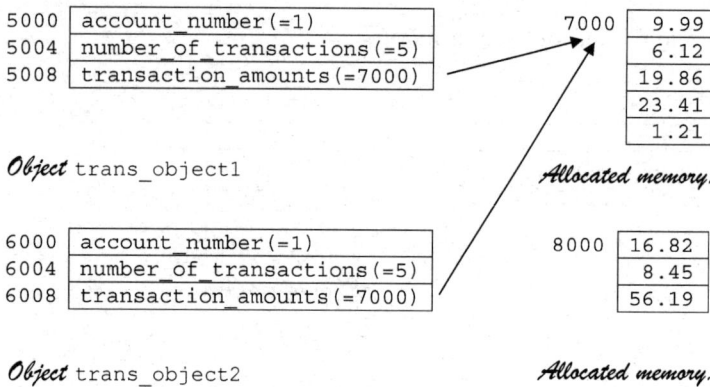

5000	account_number(=1)
5004	number_of_transactions(=5)
5008	transaction_amounts(=7000)

7000	9.99
	6.12
	19.86
	23.41
	1.21

Object trans_object1   *Allocated memory.*

6000	account_number(=1)
6004	number_of_transactions(=5)
6008	transaction_amounts(=7000)

8000	16.82
	8.45
	56.19

Object trans_object2   *Allocated memory.*

Figure 10.3

Lines 91 and 93 display the data stored in both objects. Despite having only changed the first transaction amount for trans_object1 only on line 85, the first transaction amount in trans_object2 has also changed to 9.99!

A further complication arises when the destructor on lines 40 to 43 is called for each object. At the end of the program, when the destructor de-allocates the memory pointed to by transaction_amounts for one of the objects, it also de-allocates the memory used by the other object. When the destructor is called again for the second object, it tries to de-allocate memory which has already been de-

第91行和第93行显示两个对象中存储的数据。尽管在第85行仅改变了trans_object1的第一笔交易额，但是trans_object2的第一笔交易额也被改为9.99。

更大的麻烦出现在这两个对象调用第40行至第43行的析构函数时。当程序结束时，其中一个对象调用析构函数来释放指针transaction_amounts所指向的内存区，这个操作同时也将另一个

allocated! The next program corrects theses errors by overloading the assignment operator =.

对象使用的内存区给释放了。因此，当第二个对象调用析构函数时，就发生了试图去释放已经被释放的内存区的问题。

Program Example P10K ────────────────────────────

```
1   // Program Example P10K
2   // Program to demonstrate the overloaded assignment operator =
3   // for a class with a pointer data member.
4   #include <iostream>
5   using namespace std ;
6
7   class transactions  // A class containing a pointer data member.
8   {
9   public:
10    transactions( int ac_num, int num_transactions,
11               const float transaction_values[] ) ;
12    // Purpose: Constructor.
13    ~transactions() ;
14    // Purpose: Destructor.
15    void modify_transaction( int transaction_index,float amount ) ;
16    // Purpose: Modify a transaction amount.
17    // Arguments: transaction_index - transaction to change.
18    //            amount - new transaction amount.
19    void display( void ) const ;
20    // Purpose: Display transactions.
21    const transactions& operator=( const transactions& t ) ;
22    // Purpose: Overloaded assignment operator.
23  private:
24    int account_number ;
25    int number_of_transactions ;
26    float *transaction_amounts ; // Pointer to array of transactions.
27  } ;
28
29  // Constructor.
30  transactions::transactions( int ac_num, int num_transactions,
31                          const float transaction_values[] )
32  {
33    account_number = ac_num ;
34    number_of_transactions = num_transactions ;
35    // Allocate storage for the transactions.
36    transaction_amounts = new float[number_of_transactions] ;
37    for ( int i = 0 ; i < number_of_transactions ; i++ )
38      transaction_amounts[i] = transaction_values[i] ;
39  }
40
41  // Destructor.
42  transactions::~transactions()
43  {
44    delete[] transaction_amounts ;
45  }
46
```

```
47    // Modify a transaction.
48    void transactions::modify_transaction( int transaction_index,
49                                            float amount )
50    {
51      // Modify the transaction amount if the array index is valid.
52      if ( transaction_index >= 0 &&
53            transaction_index < number_of_transactions )
54        transaction_amounts[transaction_index] = amount ;
55    }
56
57    void transactions::display( void ) const
58    {
59      cout << " Account Number " << account_number << " Transactions: ";
60      for ( int i = 0 ; i < number_of_transactions ; i++ )
61        cout << transaction_amounts[i] << ' ' ;
62      cout << endl ;
63    }
64
65    // Overloaded assignment operator.
66    const transactions& transactions::operator=(const transactions& t)
67    {
68      if ( this != &t) // Avoid self-assignment.
69      {  // Copy each member from t.
70        account_number = t.account_number ;
71        number_of_transactions = t.number_of_transactions ;
72        // Delete old array and allocate memory for new array.
73        delete[] transaction_amounts ;
74        transaction_amounts = new float[number_of_transactions] ;
75        // Copy transaction amounts from t.
76        for ( int i = 0 ; i < number_of_transactions ; i++ )
77          transaction_amounts[i] = t.transaction_amounts[i] ;
78      }
79      return *this ;
80    }
81
82    int main( void )
83    {
84      float trans_amounts1[] = { 25.67, 6.12, 19.86, 23.41, 1.21 } ;
85      transactions trans_object1( 1, 5, trans_amounts1 ) ;
86      float trans_amounts2[] = { 16.82, 8.45, 56.19 } ;
87      transactions trans_object2( 2, 3, trans_amounts2 ) ;
88
89      cout << "Before assignment:" << endl ;
90      cout << "trans_object1:" << endl ;
91      trans_object1.display() ;
92      cout << "trans_object2:" << endl ;
93      trans_object2.display() ;
94
95      trans_object2 = trans_object1 ; // The overloaded = operator is called here.
96
97      cout << endl << "After assignment:" << endl ;
```

```
98    cout << "trans_object1:" << endl ;
99    trans_object1.display() ;
100   cout << "trans_object2:" << endl ;
101   trans_object2.display() ;
102
103   // Change first transaction amount of trans_object1 to 9.99
104   trans_object1.modify_transaction( 0, 9.99 ) ;
105
106   cout << endl
107       << "After changing first transaction amount of"
108       << " trans_object1 to 9.99" << endl ;
109   cout << "trans_object1:" << endl ;
110   trans_object1.display() ;
111   cout << "trans_object2:" << endl ;
112   trans_object2.display() ;
113   return 0 ;
114 }
```

The output from this program is:

```
Before assignment:
trans_object1:
Account Number 1 Transactions: 25.67 6.12 19.86 23.41 1.21
trans_object2:
Account Number 2 Transactions: 16.82 8.45 56.19

After assignment:
trans_object1:
Account Number 1 Transactions: 25.67 6.12 19.86 23.41 1.21
trans_object2:
Account Number 1 Transactions: 25.67 6.12 19.86 23.41 1.21

After changing first transaction amount of trans_object1 to 9.99
trans_object1:
Account Number 1 Transactions: 9.99 6.12 19.86 23.41 1.21
Account Number 1 Transactions: 25.67 6.12 19.86 23.41 1.21
```

The overloaded assignment operator is called on line 95 and is equivalent to the statement:

```
trans_object2.operator=( trans_object1 ) ;
```

For efficiency purposes, line 66 declares the parameter t as a reference to a transactions object. By declaring t as a reference, the address of the object, rather than a copy of the object, is passed to t. Using const ensures that the object that t is a reference to is not changed in the function.

Line 68 checks that the calling object and the passed object are not one and the same. This is to avoid a statement such as trans_object1 = trans_object1 causing a problem. Without this check the

为了提高效率，第66行将参数t声明为一个transactions对象的引用。通过将t声明为引用，将对象的地址而并非对象的副本传递给了t。使用const可以保证t所指向的对象在函数中不会被修改。

第68行用于验证调用的对象和传递的对象不是同一个，目的是避免像trans_object1 = trans_object1这样的赋值语句所导致的错误。如果没有这项检查，第73行会将赋值运算符左右两边的同一对象（trans_object1）中为交易额所分

allocated memory for the transaction amounts of both the left hand side (`trans_object1`) and the right hand side objects (also `trans_object1`) would be simultaneously released in line 73.

Unlike the default assignment operator, the overloaded assignment operator doesn't simply copy the `transactions` pointer value. Firstly, the memory pointed by transactions is released on line 73. New memory of the correct size is then allocated and the transaction amounts are copied one by one in the loop in lines 76 and 77.

After line 87, a sketch of the memory is the same as for P10J as shown in Figure 10.1.

After line 95 a sketch of memory will look like Figure 10.4.

配的内存单元同时释放。

与默认的赋值运算符不同，重载的赋值运算符不再是对transactions指针值的简单复制。首先，第73行释放了transactions 指向的内存区。然后，大小合适的新的内存区被重新分配，并通过第76行和第77行的循环，将交易额逐个进行复制。

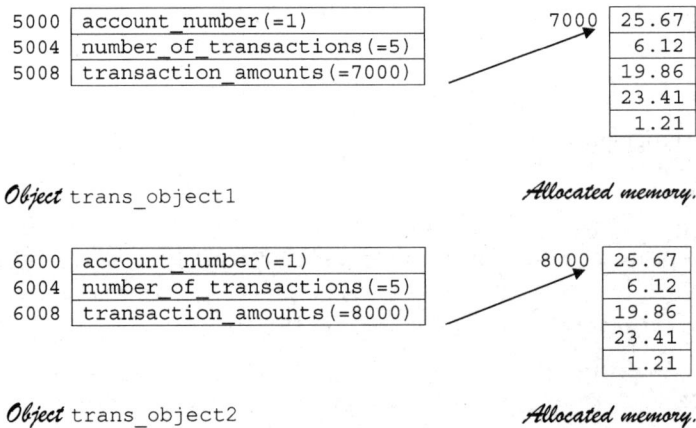

5000	account_number(=1)
5004	number_of_transactions(=5)
5008	transaction_amounts(=7000)

7000	25.67
	6.12
	19.86
	23.41
	1.21

Object trans_object1   *Allocated memory.*

6000	account_number(=1)
6004	number_of_transactions(=5)
6008	transaction_amounts(=8000)

8000	25.67
	6.12
	19.86
	23.41
	1.21

Object trans_object2   *Allocated memory.*

Figure 10.4

After changing the first transaction amount of `trans_object1` to `9.99`, a sketch of the memory will look like Figure 10.5.

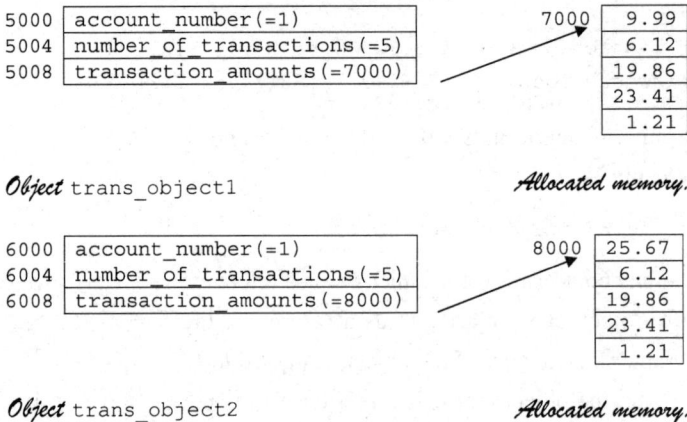

5000	account_number(=1)
5004	number_of_transactions(=5)
5008	transaction_amounts(=7000)

7000	9.99
	6.12
	19.86
	23.41
	1.21

Object trans_object1   *Allocated memory.*

6000	account_number(=1)
6004	number_of_transactions(=5)
6008	transaction_amounts(=8000)

8000	25.67
	6.12
	19.86
	23.41
	1.21

Object trans_object2   *Allocated memory.*

Figure 10.5

By comparing Figure 10.5 to Figure 10.3, it can be seen that the allocated memory for the two objects are now independent of each other.

通过比较图10.5和图10.3可以看出，为两个对象所分配的内存区是彼此独立的。

Note: The updated object is returned on line 79 to allow multiple assignments to be chained. For example, if `tr1`, `tr2` and `tr3` are `transactions` objects, the assignment operator is chained in the expression `tr1 = tr2 = tr3`. This expression is equivalent to the expression `tr1.operator=(tr2.operator=( tr3 ))`. In evaluating this expression, `tr2.operator= ( tr3 )` is evaluated first and the result is returned in a temporary variable `t`, say. The returned value `t` is then used as an argument in `tr1.operator=( t )`.

注意：函数在第79行返回的是更新后的对象，因此允许进行多重连续赋值。

10.10   The copy constructor（拷贝构造函数）

A copy constructor is a special constructor that is called automatically whenever any of the following occur:

(i) a new object is created and initialised. For example,

拷贝构造函数是一种特殊的构造函数，下列任何一种情况发生时，都会自动调用此函数。
(i) 创建并初始化一个新对象时。

```
float trans_amounts[] = { 56.88, 89.65, 15.76, 63.91 } ;
transactions trans_object1( 1, 4, trans_amounts ) ;
transactions trans_object2 = trans_object1 ;
// trans_object2 is created and initialised with
// trans_object1 member values by the copy constructor.
```

Initialisation is not the same as assignment.

The statement `transactions trans_object2 = trans_object1 ;`

in (i) above is an example of initialisation. The object `trans_object2` is created and given a value.

The statement `trans_object2 = trans_object1` is an example of an assignment. The object `trans_object2` is only given a value and is not created by this statement.

(ii) when an object is passed by value to a function. For example,

初始化与赋值是不同的。

语句transactions trans_object2 = trans_object1; 是一个初始化的例子。trans_object2对象在创建时被赋了一个值。而语句trans_object2 = trans_object1则是一个赋值的例子。对象trans_object2只是被赋了一个值，但并不是由该语句创建的。
(ii) 按传值方式将一个对象传给函数时。

```
int main( void )
{
  void any_function( transaction tr ) ;
  float trans_amounts[] = { 39.87, 6.43, 7.34, 75.12, 9.65 } ;
  transactions trans( 3, 5, trans_amounts ) ;
  any_function ( trans ) ; // trans is passed by value to
                           // the function i.e. a copy of
                           // trans is passed to tr.
  ...
  return 0 ;
}

void any_function( transactions tr )
{
  // tr receives a copy of trans.
  // The copy constructor does this.
  ...
}
```

(iii) when an object is returned by value from a function　　　　(iii) 从函数返回对象本身时。

```
int main( void )
{
  ...
  transactions another_function() ;
  ...
  trans = another_function() ; // A copy of tr is passed
                               // back to trans. The copy
                               // constructor does this.
    ...
  return 0 ;
}

transactions another_function()
{
  float trans_amounts[] = { 70.00, 68.47} ;
  transactions tr( 4, 2, trans_amounts ) ;

  ...
  return tr ; // tr is returned by value.
}
```

Each class has a default copy constructor, which, like the default assignment operator, performs a member-wise copy. For most classes the default copy constructor and the default assignment operator are adequate and work as expected. However, for classes that allocate and de-allocate memory, the default copy constructor exhibits the same problems as the default assignment operator. In general, a class that defines its own assignment operator will also have to define its own copy constructor.

The next program extends program P10K by defining a copy constructor for the transactions class.

每个类都有一个默认的拷贝构造函数，它与默认的赋值运算符类似，逐个成员进行复制。对于大多数的类而言，默认的拷贝构造函数和默认的赋值运算符都足以达到我们的要求。但是，对于需要动态分配和释放内存的类，默认的拷贝构造函数与默认的赋值运算符存在同样的问题。通常，一个定义了自己的赋值运算符的类也必须定义它自己的拷贝构造函数。

Program Example P10L

```
1   // Program Example P10L
2   // Program to demonstrate the use of the copy constructor for
3   // a class with a pointer data member.
4   #include <iostream>
5   using namespace std ;
6
7   class transactions  // A class containing a pointer data member.
8   {
9   public:
10    transactions( int ac_num, int num_transactions,
11                const float transaction_values[] ) ;
12    // Purpose: Constructor.
13    transactions( const transactions& tr ) ;
14    // Purpose: Copy Constructor.
15    ~transactions() ;
```

```
16    // Purpose: Destructor.
17    void modify_transaction( int transaction_index, float amount ) ;
18    // Purpose  : Modify a transaction amount.
19    // Arguments: transaction_index - transaction to change.
20    //              amount - new transaction amount.
21    void display( void ) const ;
22    // Purpose: Display transactions.
23    const transactions& operator=( const transactions& t ) ;
24    // Purpose: Overloaded assignment operator.
25  private:
26    int account_number ;
27    int number_of_transactions ;
28    float *transaction_amounts ; // Pointer to array of transactions.
29  } ;
30
31  // Constructor.
32  transactions::transactions( int ac_num, int num_transactions,
33                              const float transaction_values[] )
34  {
35    account_number = ac_num ;
36    number_of_transactions = num_transactions ;
37    // Allocate storage for transactions.
38    transaction_amounts = new float[number_of_transactions] ;
39    for ( int i = 0 ; i < number_of_transactions ; i++ )
40      transaction_amounts[i] = transaction_values[i] ;
41  }
42
43  // Copy constructor.
44  transactions::transactions( const transactions& tr )
45  {
46    account_number = tr.account_number ;
47    number_of_transactions = tr.number_of_transactions ;
48    transaction_amounts = new float[number_of_transactions] ;
49    for ( int i = 0 ; i < number_of_transactions ; i++ )
50      transaction_amounts[i] = tr.transaction_amounts[i] ;
51  }
52
53  // Destructor.
54  transactions::~transactions()
55  {
56    delete[] transaction_amounts ;
57  }
58
59  // Modify a transaction.
60  void transactions::modify_transaction( int transaction_index,
61                                         float amount )
62  {
63    // Modify the transaction amount if the array index is valid.
64    if ( transaction_index >= 0 &&
65         transaction_index < number_of_transactions )
66      transaction_amounts[transaction_index] = amount ;
```

```
67   }
68
69   void transactions::display( void ) const
70   {
71     cout << " Account Number " << account_number << " Transactions: " ;
72     for ( int i = 0 ; i < number_of_transactions ; i++ )
73       cout << transaction_amounts[i] << ' ' ;
74     cout << endl ;
75   }
76
77   // Overloaded assignment operator.
78   const transactions& transactions::operator=(const transactions& t)
79   {
80     if ( this != &t) // Avoid self-assignment.
81     {  // Copy each member from t.
82       account_number = t.account_number ;
83       number_of_transactions = t.number_of_transactions ;
84       // Delete old array and allocate memory for new array.
85       delete[] transaction_amounts ;
86       transaction_amounts = new float[number_of_transactions] ;
87       // Copy transaction amounts from t.
88       for ( int i = 0 ; i < number_of_transactions ; i++ )
89         transaction_amounts[i] = t.transaction_amounts[i] ;
90     }
91     return *this ;
92   }
93
94   int main( void )
95   {
96     float trans_amounts1[] = { 25.67, 6.12, 19.86, 23.41, 1.21 } ;
97     transactions trans_object1( 1, 5, trans_amounts1 ) ;
98     transactions trans_object2 = trans_object1 ;
99     // The line above calls the class copy constructor.
100    cout << "trans_object1:" << endl ;
101    trans_object1.display() ;
102    cout << "trans_object2:" << endl ;
103    trans_object2.display() ;
104    return 0 ;
105  }
```

The output from this program is:

```
trans_object1:
Account Number 1 Transactions: 25.67 6.12 19.86 23.41 1.21
trans_object2:
Account Number 1 Transactions: 25.67 6.12 19.86 23.41 1.21
```

The copy constructor for the transaction class is defined on lines 44 to 51. Like other constructors, the copy constructor has no return type, not even `void`.

与其他的构造函数相同，拷贝构造函数没有返回类型，甚至连 void 也不能有。

A copy constructor has only one parameter which is a reference to an object of the same class. The parameter is a `const` reference so that the copy constructor cannot change the object passed to it by reference. When the copy constructor is called on line 98, the parameter `tr` on line 44 becomes a reference to `trans_object1`. The non-pointer data members `account_number` and `number_of_transactions` of `trans_object2` are assigned a value by copying the corresponding members of `tr`. After memory has been allocated for the transaction amounts on line 48, the transaction amounts are copied from `tr` to `trans_object2`.

一个拷贝构造函数只能有一个形参，它是指向同一类的对象的引用。这个形参是一个const引用，拷贝构造函数不能通过这个引用来修改传递给它的对象。

10.11   Overloading the index operator [ ]
（重载下标运算符[ ]）

It is a common error to define an array with, for example, 10 elements and then attempt to access an element with an index value outside the range 0 to 9.

The next program includes a class for handling an array of integers that detects this type of error. The class contains an overloaded index operator `[ ]` for automatically checking for out-of-bounds values in the index.

经常会发生这样的错误，例如，定义了一个10个元素的数组，然后试图访问下标在0~9范围以外的数组元素。
下面的程序包含一个处理整型数组的类，用来检测此类错误。类中包含一个重载下标运算符[ ]，用于自动检测下标值越界。

Program Example P10M

```
1   // Program Example P10M
2   // Program to demonstrate the overloading of index operator [].
3   #include <iostream>
4   using namespace std ;
5
6   class int_array  // A smart integer array.
7   {
8   public:
9     int_array( int number_of_elements = 10 ) ;
10    int_array( int_array const &array ) ;
11    ~int_array() ;
12    int_array const& operator=( int_array const &array ) ;
13    int &operator[]( int index ) ;
14   private:
15    int number_of_elements ;
16    int* data ;
17    void check_index( int index) const ;
18    void copy_array( int_array const &array ) ;
19   } ;
20
21   // Constructor.
22   int_array::int_array( int n )
23   {
24     if ( n < 1 )
```

```
25    {
26      cerr << "number of elements cannot be " << n
27          << ", must be >= 1" << endl ;
28      exit( 1 ) ;
29    }
30    number_of_elements = n ;
31    data = new int [number_of_elements] ;
32    for (int i = 0 ; i < number_of_elements ; i++ )
33      data[ i ] = 0 ;
34  }
35
36  // Copy constructor.
37  int_array::int_array(int_array const &array)
38  {
39    copy_array(array) ;
40  }
41
42  // Destructor.
43  int_array::~int_array()
44  {
45    delete[] data ;
46  }
47
48  // Assignment operator.
49  int_array const& int_array::operator=( int_array const &array )
50  {
51    if ( this != &array ) // Avoid self assignment.
52    {
53      delete[] data ;
54      copy_array( array ) ;
55    }
56    return *this ;
57  }
58
59  // Overloaded index operator.
60  int& int_array::operator[]( int index )
61  {
62    check_index( index ) ;
63    return data[ index ] ;
64  }
65
66  void int_array::copy_array( int_array const &array )
67  {
68    number_of_elements = array.number_of_elements ;
69    data = new int [ number_of_elements ] ;
70    for ( int i = 0 ; i < number_of_elements ; i++ )
71      data[ i ] = array.data[ i ] ;
72  }
73
74  void int_array::check_index( int index ) const
```

```
75   {
76     if ( index < 0 || index >= number_of_elements )
77     {
78       cerr << "invalid index " << index
79            << ", range is 0 to " << number_of_elements - 1 << endl ;
80       exit( 1 ) ;
81     }
82   }
83
84   int main( void )
85   {
86     int_array a( 15 ) ; // An array of 15 integers.
87     int i ;
88
89     // Display the contents of the array.
90     for ( i = 0 ; i < 15 ; i++ )
91       cout << a[i] << ' ' ;
92     cout << endl ;
93
94     // Assign some values to the elements of the array.
95     for ( i = 0 ; i < 15 ; i++ )
96       a[i] = i * 10 ;
97
98     // Display the new contents of the array.
99     for ( i = 0 ; i < 15 ; i++ )
100       cout << a[i] << ' ' ;
101     cout << endl ;
102
103     exit( 0 ) ;
104     return 0 ;
105   }
```

The output from this program is:
```
0 0 0 0 0 0 0 0 0 0 0 0 0 0 0
0 10 20 30 40 50 60 70 80 90 100 110 120 130 140
```

The class uses a pointer to allocate the exact memory required for the size of the array specified in the constructor.

The overloaded index operator `[]` simply checks the value of the index before returning the array element. If the value of the index is invalid, an error message is displayed and the program is terminated. Lines 28 and 80 calls `exit()` to terminate the program and return to the operating system with a status code of 1. The status code is usually set to 0 to indicate a normal exit and to some other value to indicate an error.

The class also contains a copy constructor, an overloaded assignment operator and a destructor. These functions are standard for any class that manages its own memory.

在构造函数中，类使用指针为数组分配指定大小的内存空间。

重载下标运算符[ ]只是在返回数组元素前检查下标值。如果下标值无效，显示错误信息并终止程序。第28 行和第80 行调用exit( ) 来终止程序并将状态码1返回给操作系统。状态码通常设置为0，表示程序正常退出；若设置为其他值，则表示出现了错误。

这个类也包含一个拷贝构造函数、一个重载赋值运算符和一个析构函数。对于任何管理自己内存区的类而言，定义这些函数都是必需的。

Programming pitfalls

1. Overloading arithmetic operators such as + or - do not overload the corresponding compound operators += and -=. Each operator must be overloaded separately.

2. Using a class such as the `time24` class, the following will work as expected:

```
time24 t( 1, 2, 3 ) ;
++( ++t ) ;        // t is 1:02:05
```

The last statement is equivalent to:

```
( t.operator++() ).operator++() ;
```

The expression `t.operator++()` increments `t` and returns a reference to t that is used in the second call. The result is that `t` gets incremented twice.

Consider the equivalent test using the overloaded postfix operator ++.

```
time24 t( 1, 2, 3 ) ;
( t++ )++ ;        // t is now 1:02:04
```

The last statement is equivalent to:

```
( t.operator++(0) ).operator++( 0 ) ;
// 0 = dummy int argument
```

In this case, the first `t.operator++( 0 )` increments `t` and returns a copy of a temporary object (not `t`) that is used in the second call. The result is that `t` gets incremented once and the copy of the temporary object also gets incremented once.

There is no way around this problem, other than to break the expression into two separate statements: `t++ ; t++ ;`

3. If a class uses a pointer data member to point to dynamically allocated storage, the class will require an overloaded assignment operator, a destructor and a copy constructor. In general, if a class requires one of these then it will require all three.

1. 重载算术运算符（如+、-）并没有重载相应的复合运算符（如+=、-=）。如果需要，则必须对它们单独进行重载。

++(++t);等价于(t.operator++( )).operator++( );。
这里，表达式t.operator++( )先将t的值增1，然后返回t的引用而用于第二次调用，于是t的值增1两次。

(t++)++;等价于(t.operator++(0)).operator++(0);。
这里，表达式t.operator++(0)将t的值增1，同时返回一个临时对象（不是t）的副本用于第二次调用。结果是t增1了一次，临时对象的副本也增1了一次。没有其他方法能解决这个问题，除非将(t++)++;拆分成两条语句t++;和t++;才能使 t 的值增1两次。

3. 如果一个类使用了一个指针数据成员指向动态分配的内存，那么这个类必须定义一个重载赋值运算符、一个析构函数和一个拷贝构造函数。通常，如果一个类需要其中之一，那么必然会三个都需要。

Quick syntax reference

	Syntax	Examples

Operator overloading	`<return type>`
	`operator <operator symbol>`
	`( operand parameter`$_1$`,`
	`...`
	`operand parameter`$_n$` )`
	`{`
	`...`
	`}`
		`const transactions& operator=`
		`    ( const transactions& t )`
		`{`
		`...`
		`}`

Copy constructor	`class name`
	`( const class name & )`
	`{`
	`...`
	`}`
		`transactions( const transactions& t )`
		`{`
		`...`
		`}`

Summary of overloaded operator member function prototypes:

operator	parameter	returns object by	`const` function?
arithmetic (+, −, *, /)	`const` reference	by value	yes
assignment (=,+=,−=)	`const` reference	by reference	no
relational (==, !=,>,<)	`const` reference	by value	yes
prefix ++ and −−	none	by reference	no
postfix ++ and −−	`int`	by value	no
unary (−,+)	none	by value	yes
index (`[]`)	value	by reference	no

If an operator changes the value of the calling object then return a reference to the object, otherwise create a temporary object to hold the results of the operation and return the temporary object by value.

Exercises

1. Modify the `time24` class to include a constructor to initialise a `time24` object with a time in the 12-hour format. For example,

   ```
   time24 t( 1, 0, 0, "pm" ) ;
   ```

   is equivalent to:

   ```
   time24 t( 13, 0, 0 ) ;
   ```

   The prototype of the new constructor is:

   ```
   time24( int h, int m, int s, const string& am_pm ) ;
   ```

2. Modify the `time24` class to include an overloaded relational operators for less than <, greater than > and not equal to !=. See program P10F.

3. Modify the `time24` class to include an overloaded operator −=. The prototype for the new member function is:

   ```
   const time24& operator−=( const time24& t ) ;
   ```

4. A complex number k is written as k = a + bi, where a is the real part and b is the imaginary part of the number. A class to represent complex numbers will have two `private` data members.

```
class complex
{
public:
        ...
private:
        float real ;
        float imaginary ;
} ;
```

(a) define arithmetic operators + and − for complex numbers

(b) define relational operators ==, ! =, < and > for the complex numbers

(c) define input and output operators >> and << for complex numbers

(d) test the class with the following statements in `main()`:

```
complex c1, c2, c3, c4, c5 ;
cin >> c1 >> c2 ;
c3 = c1 + c2 ;
c4 = c1 - c2 ;
if ( c3 != c4 )
{
  if ( c3 < c4 )
  {
    cout << "c3 is less than c4" ;
    c5 = c4 - c3 ;
  }
  if ( c3 > c4 )
  {
    cout << "c3 is greater than c4" ;
    c5 = c3 - c4 ;
  }
}
if ( c3 == c4 )
{
  cout << "c3 and c4 are equal" ;
  c5 = c4 ;
}
cout << c5 ;
```

5. Write a class `cal_date` to represent a calendar date. The class should include the following:

(a) three `private` data members representing the day, the month and the year of a date

(b) a default constructor that initialises a date to the start of the Gregorian calendar 14/9/1752

(c) a constructor that takes three integer arguments to set a `cal_date` object to a specific date

(d) a `public` inspector member function for each of the private data members of the class

(e) a `public` mutator member function to assign a `cal_date` object to a specific date

(f) an overloaded + operator to add an integer to a `cal_date` object

(g) an overloaded – operator to subtract two cal_date objects, resulting in the number of days between the two dates

(h) overloaded ++ and −− operator functions

(i) overloaded relational operators <, >, == and !=

(j) an input operator >> and an output operator <<.

Write statements in main() to test each of the above functions.

6. Modify the int_array class of program P10M to overload the function call operator () to have the same meaning as the index operator []. This means that a[i] and a(i) will both access the same element of the integer array a.

Chapter Eleven
Inheritance
第 11 章 继　　承

11.1　What is inheritance?（什么是继承？）

Inheritance is one of the fundamental concepts of object-oriented programming. Inheritance allows a new class to be constructed using an existing class as a basis. The new class incorporates or *inherits* the data members and member functions of the existing class. Additional data members and member functions can be added to the new class, there by extending the existing class. The existing class is known as the *base* class and the new class is known as the *derived* or *inherited* class (see Figure 11.1).

继承是面向对象编程的基本概念之一。继承使我们能在已有类的基础上构造新类。这个新的类拥有或继承已有类的数据成员和成员函数，也可添加新的数据成员和成员函数，对已有的类进行扩充。这个已有的类就称为基类，而这个新类则称为派生类。

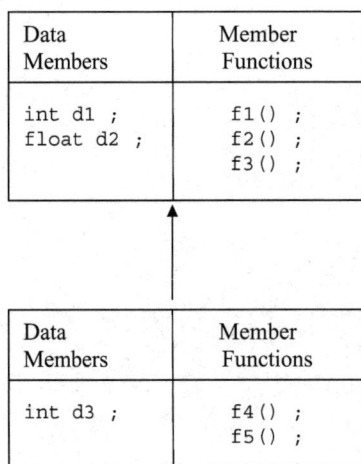

Data Members	Member Functions
int d1 ; float d2 ;	f1() ; f2() ; f3() ;

An existing class with various data members d1 *and* d2, *say. It also has member functions* f1(), f2() *and* f3().
This is called the base class.

The arrow means 'inherits from'

Data Members	Member Functions
int d3 ;	f4() ; f5() ;

A new class with an additional data member d3 *and two additional member functions* f4() *and* f5().
This class has now 3 data members and 5 member functions.
This is called the derived or inherited class.

Figure 11.1　Base and inherited class

Note that the base class is not modified in any way and is simply used as a basis for writing the derived class. This is called *reusability* (the base class is being reused) and is a major goal of object-oriented programming. Reusing an existing class that is known to be working has obvious advantages with respect to the time taken to develop and debug a new class.

The derived class can, in turn, be used as a base class from which other classes may be derived, thus creating a *class hierarchy*. For

注意，基类没有做任何改动，仅仅是作为生成派生类的基础，这称为可重用性（基类被重用），可重用性也是面向对象编程的主要目标。考虑到开发和调试一个新程序的时间开销，重用一个现成的、已知是好用的类具有明显的优势。

派生类也能作为基类，再派生出其他的类，这样就形成了一个

example, a factory, a college and a hospital are all specific examples of a more general concept known as a building. Every building has a basic set of properties such as the floor area, the number of windows, the number of doors, etc. A college building, for example, has some additional properties such as a number of lecture theatres and computer laboratories. A hospital has operating theatres, wards, etc. These properties are unique to these buildings and do not, in general, exist in all buildings. A factory will have yet another set of additional properties that are unique to factories.

These types of buildings can be further sub-divided into different types of factories, colleges and hospitals.

Using buildings as an example, a simple class hierarchy would look like Figure 11.2.

类的层次结构。例如，工厂、大学、医院都是"建筑物"这个一般概念的特例。每一个建筑物都会有一些固有的基本属性——房屋面积，窗户的数量、门的数量，等等。但是一些建筑物，比如说大学，会有一些额外的属性，如一些演讲礼堂和计算机实验室；而医院会有手术室、病房等。这些属性只属于这些建筑物，而且一般来说并不是在所有的建筑物中都存在。工厂也会有一些额外的特有属性。这些建筑类型可以进一步细分为工厂、大学和医院等不同类型。

以建筑物为例，一个简单的类层次结构如下。

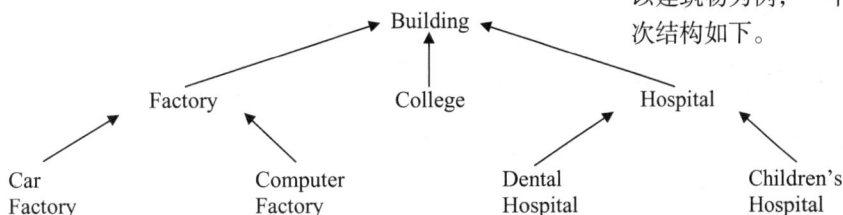

Figure 11.2　Building hierarchy

Moving down the hierarchy from the top to the bottom, the classes go from a general base class to more specialised classes that have been extended from the base class.

This type of relationship between classes is known as "*a-kind-of*" relationship. For example, a car factory is a kind of factory and a factory is a kind of building.

There are plenty of other examples of inheritance in common objects that we see or use in everyday life. Mountain bicycles and racing bicycles are specialisations of bicycles, personal computers and mainframe computers are specialisations of computers, books on object-oriented programming are specialisations of books on programming which are a specialisation of books.

在图11.2中，类随着层次结构的自顶向下，逐渐从一般基类变成更加具体的从基类扩展出来的子类。

这种类与类之间的关系称为"a-kind-of（是一种……）"关系。

11.2　Inheritance syntax（继承语法）

In its simplest form, inheritance is used by first defining the base class:

首先定义基类，然后再定义派生类。

```
class base
{
// class data members and
// class member functions.
} ;
```

This is the same as defining any class.

Next the derived classes are defined:

Note the keyword public.

```
class inherited : public base
{
// additional class data members for this class and
// additional class member functions for this class.
} ;
```

The next program implements part of the buildings class hierarchy, by defining a base class for a building and deriving a college class from this base class.

Program Example P11A

```
1  // Program Example P11A
2  // Simple demonstration of inheritance.
3  #include<iostream>
4  using namespace std ;
5
6  class building  // This is the base class.
7  {
8  public:
9    building() ;
10   float get_floor_area( void ) ;
11   int get_number_of_windows( void ) ;
12 private:
13   float floor_area ; // Some simple properties of a building.
14   int number_of_windows ;
15 } ;
16
17 // Base class member functions.
18 building::building()
19 {
20   cout << "Enter floor area: " ;
21   cin >> floor_area ;
22   cout << "Enter number of windows: " ;
23   cin >> number_of_windows ;
24 }
25
26 float building::get_floor_area( void )
27 {
28   return floor_area ;
29 }
30
31 int building::get_number_of_windows( void )
32 {
33   return number_of_windows ;
34 }
35
36 class college : public building  // This is the derived class.
```

```
37 {
38 public:
39   college() ;
40   int get_number_of_lecture_theatres( void ) ;
41   int get_number_of_laboratories( void ) ;
42 private:
43   int number_of_lecture_theatres ; // Some simple properties.
44   int number_of_laboratories ;
45 } ;
46
47 // Derived class member functions.
48 college::college()
49 {
50   cout << "Enter number of lecture theatres: " ;
51   cin >> number_of_lecture_theatres ;
52   cout << "Enter number of laboratories: " ;
53   cin >> number_of_laboratories ;
54 }
55
56 int college::get_number_of_lecture_theatres( void )
57 {
58   return number_of_lecture_theatres ;
59 }
60
61 int college::get_number_of_laboratories( void )
62 {
63   return number_of_laboratories ;
64 }
65
66 int main( void )
67 {
68   college my_college ;
69
70   // Display the properties derived from the building class.
71   cout << endl << "Floor area = " ;
72   cout << my_college.get_floor_area() << endl ;
73   cout << "Number of windows = " ;
74   cout << my_college.get_number_of_windows() << endl ;
75
76  // Display the properties specific to the college class.
77   cout << "Number of lecture theatres = " ;
78   cout << my_college.get_number_of_lecture_theatres() << endl;
79   cout << "Number of laboratories = " ;
80   cout << my_college.get_number_of_laboratories() << endl ;
81   return 0 ;
82 }
```

A sample run of this program is:

```
Enter floor area: 1000
Enter number of windows: 2
```

```
Enter number of lecture theatres: 1
Enter number of laboratories: 4

Floor area = 1000
Number of windows = 2
Number of lecture theatres = 1
Number of laboratories = 4
```

Lines 6 to 34 declare the base class `building`. This is declared just like any other C++ class. Lines 36 to 64 declare the inherited class `college`.

When the object `my_college` is defined on line 68, the constructor for the base class (lines 18 to 24) is called first, followed by the constructor for the derived class (lines 48 to 54).

The constructor for the base class is always called before the constructor of a derived class. The opposite applies in the case of destructors; the destructor of the base class is always called after the destructor of the derived class.

Lines 71 to 80 display the data members of `my_college`, which include the data members inherited from the base class. The member functions of the base class are also inherited as demonstrated on line 72. The function `get_floor_area()` is a member of the base class `building`, but is used by the college object `my_college` to display its floor area.

To further demonstrate inheritance, the next example uses a simplified class hierarchy to represent current and past employees of an organisation (see Figure 11.3).

基类的构造函数总是在派生类的构造函数被调用之前被调用。然而析构函数的调用次序正好相反：基类的析构函数总是在派生类的析构函数被调用之后被调用。第71行至第80行显示了my_college的数据成员，其中也包括从基类继承的数据成员。第72行表明基类的成员函数也被派生类继承了。get_floor_ area( )是基类building的成员函数，但是college类的对象my_college同样可以利用它来显示自己的房屋面积。

为了进一步阐述继承，下面的例子使用一个简化的类层次结构来表示一个公司目前和以前的雇员。

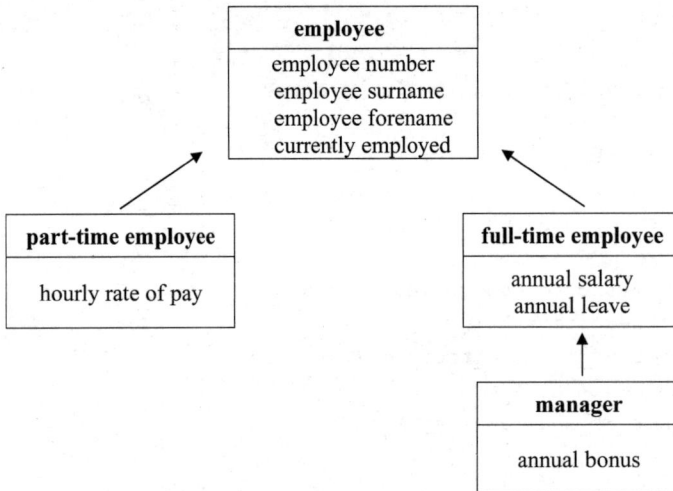

Figure 11.3　Employee hierarchy

Employees in the company are either part-time or full-time. The employee number and name are common to both types of employee. Whether the person is currently employed or is a past employee is also common to both types of employees.

Part-time employees are paid an hourly rate while full-time employees are on a salary and are allowed an annual paid leave. Managers are also full-time employees, who in addition to their annual salary and paid leave are also paid an annual bonus.

公司里的雇员分为兼职和全职两类。这两类雇员都有雇员编号和雇员名字，而且都有目前雇用的雇员或者是过去曾经雇用的雇员。兼职雇员按时薪领取工资，而全职雇员则领取月薪，并且每年还有一次带薪休假。管理者也是全职雇员，除了年薪和年度带薪休假，他们还有年度奖金。

Program Example P11B

```
1    // Program Example P11B
2    // Demonstration of inheritance.
3    #include<iostream>
4    #include<string>
5    using namespace std ;
6
7    class employee // Base class.
8    {
9    public:
10     employee() ;
11     void display_data( void ) ;
12     void left( void ) ;
13   private:
14     unsigned int employee_number ;
15     string surname ;
16     string forename ;
17     bool currently_employed ;
18   } ;
19
20   // employee member functions.
21   employee::employee()
22   {
23     cout << endl << "Enter Employee Number: " ;
24     cin >> employee_number ;
25     cout << "Enter Employee Name: " ;
26     cin >> surname >> forename ;
27     currently_employed = true ;
28   }
29
30   void employee::left( void )
31   {
32     currently_employed = false ;
33   }
34
35   void employee::display_data( void )
36   {
37     if ( currently_employed )
38       cout << "Currently Employed" ;
39     else
```

```
40        cout << "Not Currently Employed" ;
41     cout << endl << "Employee Number: " << employee_number << endl
42            << "Name: " << surname << ' ' << forename << endl ;
43  }
44
45  class part_time : public employee
46  { // part_time is a kind of employee.
47  public:
48     part_time() ;
49     void display_data( void ) ;
50  private:
51     double hourly_rate ;
52  } ;
53
54  // part_time member functions.
55  part_time::part_time()
56  {
57     cout << "Enter Hourly Rate: " ;
58     cin >> hourly_rate ;
59  }
60
61  void part_time::display_data( void )
62  {
63     employee::display_data() ;
64     cout << "Hourly Rate: " << hourly_rate << endl ;
65  }
66
67  class full_time : public employee
68  {  // full_time is a kind of employee.
69  public:
70     full_time() ;
71     void display_data( void ) ;
72  private:
73     double annual_salary ;
74     int annual_leave ;
75  } ;
76
77  // full_time member functions.
78  full_time::full_time()
79  {
80     cout << "Enter Salary: " ;
81     cin >> annual_salary ;
82     cout << "Enter Annual Leave (in days): " ;
83     cin >> annual_leave ;
84  }
85
86  void full_time::display_data( void )
87  {
88     employee::display_data() ;
89     cout << "Salary: " << annual_salary << endl ;
```

```
90      cout << "Annual Leave: " << annual_leave << endl ;
91   }
92
93   class manager : public full_time
94   { // manager is a kind of full_time employee.
95   public:
96      manager() ;
97      void display_data( void ) ;
98   private:
99      double bonus ;
100  } ;
101
102  // manager member functions.
103  manager::manager()
104  {
105     cout << "Enter Bonus: " ;
106     cin >> bonus ;
107  }
108
109  void manager::display_data( void )
110  {
111     full_time::display_data() ;
112     cout << "Bonus: " << bonus << endl ;
113  }
114
115  int main( void )
116  {
117     part_time pt ;
118     full_time ft ;
119     manager man ;
120
121     // Display employee data.
122     cout << endl << "Part-time Employee Data:" << endl ;
123     pt.display_data() ;
124     cout << endl<< "Full-time Employee Data:" << endl ;
125     ft.display_data() ;
126     man.left() ;
127     cout << endl << "Manager Employee Data:" << endl ;
128     man.display_data() ;
129     return 0 ;
130  }
```

A sample run of this program is:

```
Enter Employee Number: 123
Enter Employee Name: Smith John
Enter Hourly Rate: 5.12

Enter Employee Number: 124
Enter Employee Name: Jones Mary
Enter Salary: 21500
Enter Annual Leave (in days): 21
```

```
Enter Employee Number: 125
Enter Employee Name: Other A.N.
Enter Salary: 32000
Enter Annual Leave (in days): 30
Enter Bonus: 9500

Part-time Employee Data:
Currently Employed
Employee Number: 123
Name: Smith John
Hourly Rate: 5.12

Full-time Employee Data:
Currently Employed
Employee Number: 124
Name: Jones Mary
Salary: 21500
Annual Leave: 21

Manager Employee Data:
Not Currently Employed
Employee Number: 125
Name: Other A.N.
Salary: 32000
Annual Leave: 30
Bonus: 9500
```

The derived classes `part_time` and `full_time` inherit the data members `employee_number`, `surname`, `forename` and `currently_employed` from the base class employee. The manager class also derives `annual_salary` and `annual_leave` from its base class `full_time`, which in turn is derived from the `employee` class. The class manager is said to be *indirectly derived* from the class `employee` and *directly derived* from the class `full_time`.

The member functions of `employee` are also inherited as shown by the use of the base class member function `left()` by the manager object `man` on line 126.

A derived class can have a data member or member function with the same name as one in its base class. For example, each of the derived classes has a member function `display_data()`. In this case, the base class member function `display_data()` is *overridden* by the inherited class member function of the same name. However, the base class member function `display_data()` can still be called to display the base class data members by the statement:

派生类可以拥有和基类同名的数据成员和成员函数。这时，基类的成员函数display_data( )将被派生类中的同名函数所屏蔽。

```
employee::display_data() ;
```

This is shown on lines 63 and 88.

The manager class has its own `display_data()` member function which calls (line 111) the `display_data()` member function of the `full_time` class, which in turn calls (line 88) the `display_data()` member function of the `employee` class. Note the use of the scope resolution operator (::) to remove any ambiguity as to which `display_data()` function to call.

注意作用域运算符（::）的使用，利用它可以消除调用display_data()时引起的歧义。

11.3   Passing arguments to a base class constructor
（向基类的构造函数传递实参）

Program P11B used constructors to initialise the data members from keyboard input. To allow for initialisation of an object when it is created, a constructor with a parameter list is required.

When a derived object is created, it should take care of the construction of the base object by calling the constructor for the base class. As seen in the previous program, the base class default constructor is automatically called from the derived class. When an object has initial values, the initialisation of the base class object is done by the use of an initialisation list in the derived class constructor.

The next program demonstrates the technique.

为了让一个对象在创建时就被初始化，需要使用一个带有形参列表的构造函数。

在创建一个派生类的对象时，应注意基类也会调用它的构造函数去构造一个基类对象。就像前面的程序里那样，基类的默认构造函数会在派生类中自动调用。当派生类的对象有初始化值时，基类对象的初始化会通过派生类构造函数中的初始化列表完成。

Program Example P11C

```
1   // Program Example P11C
2   // Demonstration of inheritance.
3   #include<iostream>
4   #include<string>
5   using namespace std ;
6
7   class employee // Base class.
8   {
9   public:
10    employee() ;
11    employee( unsigned int number,
12              string sname, string fname ) ;
13    void display_data( void ) ;
14    void left( void ) ;
15  private:
16    unsigned int employee_number ;
17    string surname ;
18    string forename ;
19    bool currently_employed ;
20  } ;
21
22  // employee member functions.
23  employee::employee()
24  {
25    cout << endl << "Enter Employee Number: " ;
```

```
26     cin >> employee_number ;
27     cout << "Enter Employee Name: " ;
28     cin >> surname >> forename ;
29     currently_employed = true ;
30   }
31
32   employee::employee( unsigned int number,
33                       string sname, string fname )
34   {
35     employee_number = number ;
36     currently_employed = true ;
37     surname = sname ;
38     forename = fname ;
39   }
40
41   void employee::left( void )
42   {
43     currently_employed = false ;
44   }
45
46   void employee::display_data( void )
47   {
48     if ( currently_employed )
49        cout << "Currently Employed" ;
50     else
51        cout << "Not Currently Employed" ;
52     cout << endl << "Employee Number: " << employee_number << endl
53          << "Name: " << surname << ' ' << forename << endl ;
54   }
55
56   class part_time : public employee
57   {  // part_time is a kind of employee.
58   public:
59     part_time() ;
60     part_time( unsigned int number,
61                string sname, string fname,
62                double rate ) ;
63     void display_data( void ) ;
64   private:
65     double hourly_rate ;
66   } ;
67
68   // part_time member functions.
69   part_time::part_time()
70   {
71     cout << "Enter Hourly Rate: " ;
72     cin >> hourly_rate ;
73   }
74
75   part_time::part_time( unsigned int number,
```

```
76                      string sname, string fname,
77                      double rate )
78            : employee( number, sname, fname ), hourly_rate( rate )
79  {}
80
81  void part_time::display_data( void )
82  {
83    employee::display_data() ;
84    cout << "Hourly Rate: " << hourly_rate << endl ;
85  }
86
87  class full_time : public employee
88  {  // full_time is a kind of employee.
89  public:
90    full_time() ;
91    full_time( unsigned int number,
92              string sname, string fname,
93              double salary, int leave ) ;
94    void display_data() ;
95  private:
96    double annual_salary ;
97    int annual_leave ;
98  } ;
99
100  // full_time member functions.
101  full_time::full_time()
102  {
103    cout << "Enter Salary: " ;
104    cin >> annual_salary ;
105    cout << "Enter Annual Leave (in days): " ;
106    cin >> annual_leave ;
107  }
108
109  full_time::full_time( unsigned int number,
110                        string sname, string fname,
111                        double salary, int leave )
112            : employee( number, sname, fname ),
113              annual_salary( salary ), annual_leave( leave )
114  {}
115
116  void full_time::display_data( void )
117  {
118    employee::display_data() ;
119    cout << "Salary: " << annual_salary << endl ;
120    cout << "Annual Leave: " << annual_leave << endl ;
121  }
122
123  class manager : public full_time
124  {  // manager is a kind of full_time employee.
125  public:
```

```
126   manager() ;
127   manager( unsigned int number,
128            string sname, string fname,
129            double salary, int leave, double bonus ) ;
130   void display_data( void ) ;
131 private:
132   double bonus ;
133 } ;
134
135   // manager member functions.
136 manager::manager()
137 {
138   cout << "Enter Bonus: " ;
139   cin >> bonus ;
140 }
141
142 manager::manager( unsigned int number,
143                   string sname, string fname,
144                   double salary, int leave, double annual_bonus )
145          : full_time( number, sname, fname, salary, leave ),
146            bonus( annual_bonus )
147 {}
148
149 void manager::display_data( void )
150 {
151   full_time::display_data() ;
152   cout << " Bonus: " << bonus << endl ;
153 }
154
155 int main( void )
156 {
157   part_time pt( 123, "Smith", "John", 5.12 ) ;
158   full_time ft( 124, "Jones", "Mary", 21500, 21 ) ;
159   manager man( 125, "Other", "A.N.", 32000, 30, 9500 ) ;
160
161   // Display employee data.
162   cout << endl << "Part-time Employee Data:" << endl ;
163   pt.display_data() ;
164   cout << endl << "Full-time Employee Data:" << endl ;
165   ft.display_data() ;
166   man.left() ;
167   cout << endl << "Manager Employee Data:" << endl ;
168   man.display_data() ;
169   return 0 ;
170 }
```

The output from this program is:

```
Part-time Employee Data:
Currently Employed
Employee Number: 123
Name: Smith John
```

```
Hourly Rate: 5.12

Full-time Employee Data:
Currently Employed
Employee Number: 124
Name: Jones Mary
Salary: 21500
Annual Leave: 21

Manager Employee Data:
Not Currently Employed
Employee Number: 125
Name: Other A.N.
Salary: 32000
Annual Leave: 30
Bonus: 9500
```

Line 157 now initialises the object `pt` with the values in the parentheses. The constructor on lines 75 to 79 is called. The expression `employee ( number, sname, fname )` in the initialisation list on line 78 is an explicit call to the base class constructor on lines 32 to 39. The base class constructor is always called to initialise the base class data members before the data members of the derived class are initialised. The actual order of the items in the initialisation list is not important. The constructor on lines 75 to 79 can therefore be also written as:

基类总是在派生类初始化数据成员之前，调用自己的构造函数来初始化基类数据成员。初始化列表中的各个数据项的实际顺序并不重要。

```
part_time::part_time( unsigned int number,
                string sname, string fname,
                double rate )
        : hourly_rate( rate ), employee( number, sname, fname )
{ }
```

11.4 Protected class members（受保护的类成员）

The keywords `private` and `public` are used to control access to both the data members and the member functions of a class. A private data member or member function of a base class cannot be accessed from a member function of a derived class, thus ensuring that the principle of data hiding is upheld.

In practice it can be convenient to share data between a base class and a derived class. There are two ways of doing this. One possibility is to change the access level of the relevant `private` members to `public`, thus allowing the derived class the required access to these members.

However, this violates the principle of data hiding by making the data members available for uncontrolled modification from any part of the program. To get over this problem, C++ has a third level of

关键字private和 public用来控制对类的数据成员和成员函数的访问。派生类的成员函数不能访问基类的私有数据成员和成员函数，从而保证了信息隐藏的原则。实际中，在基类和派生类之间共享数据会更方便一些。有两种方法可以实现这个目标。一种是把相关的private成员的访问控制级别修改为public，这样派生类就可以访问这些成员了。

但是这样导致数据可以被程序中的任何部分随意地访问和修改，这又违背了信息隐藏的原则。

access known as `protected` access. This level of access allows derived classes (and only derived classes) to have access to specified base class members.

The next program demonstrates the use of `protected` class data members.

为了解决这个问题，C++提供了第三种访问控制级别protected，它允许派生类（并且只允许派生类）访问被指定为protected的基类成员。

Program Example P11D

```
1  // Program Example P11D
2  // Demonstration of protected class data members.
3  #include<iostream>
4  using namespace std ;
5
6  class rectangle
7  {
8  public:
9     rectangle( int w, int h ) ;
10    int calc_area( void ) ;
11    void display_dimensions( void ) ;
12    void display_area( void ) ;
13 protected:     // Available to derived classes.
14    int width, height ;
15 };
16
17 // rectangle member functions.
18 rectangle::rectangle( int w, int h ) : width( w ), height( h )
19 {}
20
21 int rectangle::calc_area( void )
22 {
23   return width * height ;
24 }
25
26 void rectangle::display_dimensions( void )
27 {
28   cout << "Dimensions of rectangle: " << width
29        << " X " << height << endl ;
30 }
31
32 void rectangle::display_area( void )
33 {
34   cout << "Area of rectangle: " << calc_area() << endl ;
35 }
36
37 class square : public rectangle
38 {
39 public:
40   square( int size ) ;
41   void display_dimension( void ) ;
42   void display_area( void ) ;
```

```
43 } ;
44
45 // square member functions.
46 square::square( int size ) : rectangle( size, size )
47 {}
48
49 void square::display_dimension( void )
50 {
51   cout << "Dimension of square: " << width << endl ;
52 }
53
54 void square::display_area( void )
55 {
56   cout << "Area of square: " << calc_area() << endl ;
57 }
58
59 int main( void )
60 {
61   rectangle r( 1, 2 ) ;
62   square s( 3 ) ;
63
64   r.display_dimensions() ;
65   s.display_dimension() ;
66   r.display_area() ;
67   s.display_area() ;
68   return 0 ;
69 }
```

The output from this program is:

```
Dimensions of rectangle: 1 X 2
Dimension of square: 3
Area of rectangle: 2
Area of square: 9
```

In this example, the class `square` (lines 37 to 57) is derived from a rectangle class (lines 6 to 35).

Line 61 defines a `rectangle` object `r` with sides of width 1 and height 2.

Line 62 defines a `square` object s with sides of size 3.

Line 64 displays the dimensions of the rectangle and line 65 displays the dimension of the square. In order to display the dimension of the square, line 51 requires access to the base class data member `width`. This is achieved by making `width` a `protected` data member by placing it in the `protected` section of the class.

It could be argued that `width` should be a `private` data member and an inspector member function should be included in the base class to return its value. However, accessing data members through inspector functions is less efficient than accessing them directly. Where efficiency

也许有人会说，应该将width设置为private类型的数据成员，并且基类应该有一个检测函数来返回它的值。但是通过检测函数访问

is critical, `protected` base class members provide an alternative to the overheads involved in calling inspector functions.

In summary, if a base class member is declared as `private`, then this member is not accessible outside the base class, not even from a derived class. Protected members are accessible from within a base class and from within any of its derived classes. Public members (usually member functions) are accessible from any part of the program (see Figure 11.4).

数据成员不如直接访问它们效率高。当效率至关重要时，将基类成员声明为protected类型，可以降低因调用检测函数而引起的开销。

总之，如果基类成员被声明为private类型，那么在基类之外都不能访问该成员，即使在派生类中也不能访问。protected成员在基类和基类的任何一个派生类中都可以被访问。public成员（通常是成员函数）在程序的任何部分都可以被访问。

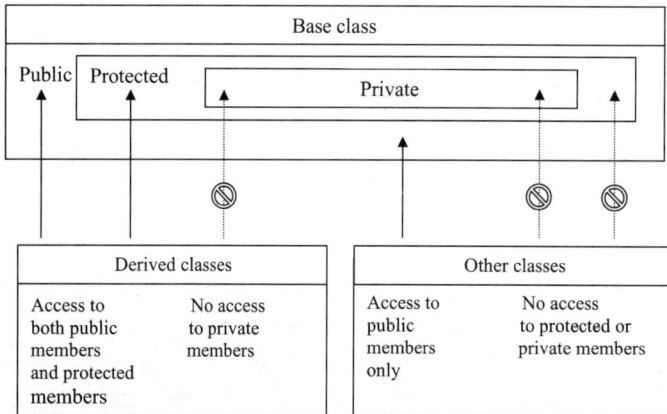

Figure 11.4   Private, protected and public members

11.5   Types of inheritance: `public`, `protected` and `private`
（继承的类型：public、protected 和private）

Declaring a base class member as `public`, `protected` or `private` is not the only way of setting the access rights of a base class member within a derived class. The access rights of a base class member within a derived class can also be specified when the derived class is declared. In the previous examples, `public` inheritance was used. For example line 37 of P11D defines the inherited class `square` with:

```
class square : public rectangle
```

The keyword `public` defines the type of inheritance to be applied to the derived class.

Public inheritance specifies that the access rights of the inherited members will be the same as they were in the base class.

Replacing the keyword `public` with the keyword `protected` makes all `public` members in the base class become `protected` members in the derived class; `protected` members in the base class remain `protected` members in the derived class.

Private inheritance is specified by using the keyword `private`. With `private` inheritance, all the `public` and `protected` members

声明基类成员为public、protected 或者 private并不是设置派生类中基类成员访问权限的唯一办法。派生类中基类成员的访问权限也可以在声明派生类时指定。

关键字public定义了应用于派生类的继承类型。

公有继承指定被继承的成员在派生类中拥有和在基类中相同的访问权限。

如果使用关键字 protected 替换 public，那么基类中所有的public成员在派生类中都会变成protected成员，而基类中的protected成员在派生类中仍为protected成员。

私有继承是用关键字private指定的。使用私有继承时，基类中的所有public和protected成员 在派

of the base class become `private` in the inherited class.

In all cases, whether it is `public`, `protected` or `private` inheritance, `private` members in a base class are inaccessible in a derived class.

The three types of inheritance are summarised in the following Table 11.1.

生类中都将变成private成员。无论是public、protected或private继承，基类的private成员在派生类中都是不可访问的。

Table 11.1   Summary of inheritance types

Type of inheritance	Base class member access	Derived class member access
public	public	public
	protected	protected
	private	inaccessible
protected	public	protected
	protected	protected
	private	inaccessible
private	public	private
	protected	private
	private	inaccessible

Public inheritance is almost always used in practice because it models "*a-kind-of*" relationship found in inheritance.

Private inheritance is used much less frequently than `public` inheritance; it can be used when none of the members of the base class is used in the derived class or by objects created from the derived class. Protected inheritance is used when the members of the base class are used in the derived class but are not required by objects created from the derived class.

As an example of how to use `protected` inheritance, consider program P11D. What would happen if a `square` object such as `s` used the `rectangle` member function `display_dimensions()`? A statement such as:

```
s.display_dimensions()
```

would display:

```
Dimensions of rectangle: 3 X 3
```

The base class member function `display_dimensions()` is not applicable to objects of the derived class `square`. One way to prevent a `square` object from using the base class member function `display_dimensions()` is to write a 'dummy' member function `display_dimensions()` in the derived class, which will simply display an error message. This will override the base class function of the same name.

A preferred approach is to make the base class function unavailable to objects of the derived class by changing the inheritance type from `public` to `protected`. The statement `s.display_dimensions()` will then generate a compiler error.

公有继承模拟了继承中的泛化关系（父类与子类之间的关系），所以在实践中应用较多。

与公有继承相比，私有继承用得较少；只有当派生类或派生类的对象不需要使用基类的任何成员时，才会使用私有继承。

当基类中的成员在派生类中使用而不在派生类的对象中使用时，才使用protected继承。

可见，基类的成员函数display_dimensions( )并不适用于派生类square 的对象。为了防止square对象调用基类成员函数display_dimensions( )，一种方法是在派生类中实现一个仅显示一条错误信息的哑元成员函数display_dimensions( )。它将屏蔽掉基类中的同名函数。使基类成员函数不被派生类对象访问的一种更好的方法是，将继承类型由public改为protected。

11.6　Composition（组合）

Inheritance depicts "a-kind-of " relationship between classes. The other type of relationship that may exist between classes is the "*has-a*" relationship, also known as *composition*. In C++ this means that objects of one class are composed of objects of another class.

The next program uses a simple class point used to represent the row and column position of a point on a screen. From this class a line class is constructed using composition.

继承描述了类与类之间的 "a-kind-of" 关系。类与类之间还存在另外一种关系——"has-a"，又被称为组合关系。在C++中这意味着一个类的对象由其他类的对象组成。

Program Example P11E

```
1   // Program Example P11E
2   // Demonstration of composition in C++.
3   #include <iostream>
4   using namespace std ;
5
6   class point // A simple point class.
7   {
8   public:
9     point() ;
10    point( int row, int col ) ;
11    int get_x( void ) ;
12    int get_y( void ) ;
13  private:
14    int x, y ; // x and y coordinates of a point.
15  } ;
16
17  // point member functions.
18  point::point()
19  {
20    x = 0 ; y = 0 ;
21  }
22
23  point::point( int row, int col ) : x( row ), y( col )
24  {}
25
26  int point::get_x( void )
27  {
28    return x ;
29  }
30
31  int point::get_y( void )
32  {
33    return y ;
34  }
35
36  class line
37  {
```

```
38 public:
39   line( point start, point end ) ;
40   float slope( void ) ;                    A line is composed of two joined points.
41 private:
42   point start, end ;   // Starting and ending points of a line.
43 } ;
44
45 // line member functions.
46 line::line( point start_point, point end_point )
47      : start( start_point ), end( end_point )
48 {}
49
50 float line::slope( void ) // Calculate the slope of a line.
51 {
52   int x1, y1, x2, y2 ;
53   float slope ;
54
55   x1 = start.get_x() ;
56   y1 = start.get_y() ;
57   x2 = end.get_x() ;
58   y2 = end.get_y() ;
59   slope = static_cast<float>(y2 - y1) /  (x2 - x1) ;
60   return slope ;
61 }
62
63 int main( void )
64 {
65   point start( 10, 20 ), end( 30, 70 ) ;
66
67   line L( start, end ) ;
68   cout << "Slope of line L is " << L.slope() << endl ;
69   return 0 ;
70 }
```

The output from this program is:

```
Slope of line L is 2.5
```

A line is composed of two point objects. When a line object is created, the constructor for the point object members is called before the constructor for the line object is called.

Whereas inheritance defines a relationship between classes, composition defines a relationship between objects.

11.7   Multiple inheritance（多重继承）

C++ allows inheritance from more than just one base class. As shown in Figure 11.5, this allows the construction of a hierarchy to express a relationship such as, class c is "a-kind-of" class a and is also "a-kind-of" class b.

一个line对象是由两个point对象构成的。当创建一个line对象时，point对象成员的构造函数总是先于line对象的构造函数被调用。继承定义的是类之间的关系；而组合定义的是对象之间的关系。

C++允许多个类作为基类，从而允许构造如图11.5所示的层次结构。这种层级结构既能表示类c是类a的子类，也能表示类c是类b的子类。

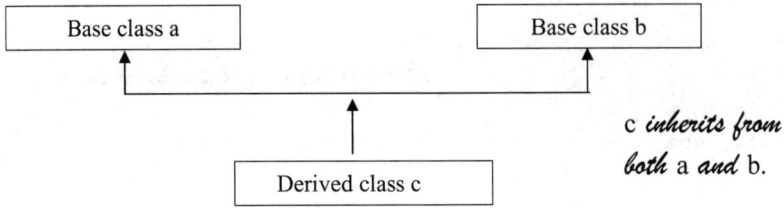

Figure 11.5　Example for inheritance from more than one base class

Some example of multiple inheritance are:

- A clock radio is a kind of clock and is also a kind of radio.
- A fax-modem is a kind of fax and is also a kind of modem.
- A seaplane is a kind of plane and is also a kind of boat.

The next program uses multiple inheritance to derive a class date_ time from existing date and time classes. The date class is called calendar and it is declared in the file calendar.h:

```
// Header for a very simple date class.
#if !defined calendar_H
#define calendar_H

class calendar
{
public:
  calendar( int d = 1, int m = 1, int y = 1970 ) ;
  void set_date( int d, int m, int y ) ;
  void get_date( int& d, int& m, int& y ) const ;
protected:
  int day ;
  int month ;
  int year ;
} ;

#endif
```

The calendar class member functions are contained in the file calendar.cpp.

```
#include <iostream>
#include <iomanip>
#include "calendar.h"
using namespace std ;
// Constructor.
calendar::calendar( int d, int m, int y )
          : day( d ), month( m ), year( y )
{}

// Mutator function.
void calendar::set_date( int d, int m, int y )
```

```
{
  day = d ; month = m ; year = y ;
}


// Inspector function.
void calendar::get_date( int& d, int& m, int& y ) const
{
  d = day ; m = month ; y = year ;
}
```

The time class is the `time24` class from program P10H. The class
header is in the file `time24.h` and contains lines 7 to 23 of program
P10H. The `private` data members have been changed to `protected`
so that they can be used by inherited classes. The `time24` class
member functions are in the file `time24.cpp`.

Program Example P11F

```
1   // Program Example P11F
2   // Demonstration of multiple inheritance.
3   #include <iostream>
4   #include "time24.h"
5   #include "time24.cpp"
6   #include "calendar.h"
7   #include "calendar.cpp"
8   using namespace std ;
9
10  class date_time : public calendar, public time24
11  {
12  public:
13    date_time( int d = 1, int mon = 1, int y = 1970,
14               int h = 0, int min = 0, int s = 0 ) ;
15  } ;
16
17  date_time::date_time( int d, int mon, int y,
18                        int h, int min, int s )
19           : calendar( d, mon, y ), time24( h, min, s )
20  {}
21
22  int main( void )
23  {
24    date_time dt( 25, 12, 1999, 0, 0, 0 ) ;
25    int day, month, year, hours, mins, secs ;
26
27    dt.get_date( day, month, year ) ;
28    dt.get_time( hours, mins, secs ) ;
29
30    cout << "Date and time is:" << endl ;
31    cout << day << "/" << month << "/" << year << endl ;
32    cout << hours << ":" << mins << ":" << secs << endl ;
33
```

```
34    dt.set_date( 1, 1, 2000 ) ;
35    dt.set_time( 1, 2, 3 ) ;
36
37    dt.get_date( day, month, year ) ;
38    dt.get_time( hours, mins, secs ) ;
39
40    cout << "Date and time is now:" << endl ;
41    cout << day << "/" << month << "/" << year << endl ;
42    cout << hours << ":" << mins << ":" << secs << endl ;
43    return 0 ;
44  }
```

The output from this program is:

```
Date and time is:
25/12/1999
0:0:0
Date and time is now:
1/1/2000
1:2:3
```

Lines 10 to 20 define a date_time class, derived from the two base classes calendar and time24. The date_time object dt defined on line 24 has six data members: day, month, year, hours, minutes and seconds inherited from the two base classes.

Lines 27 and 28 get the time and date values of dt using the member functions get_date() and get_time(), inherited from the calendar and time24 base classes.

Similarly, the date and time values of dt are assigned values on lines 34 and 35 using the member functions set_date() and set_time(), inherited from the calendar and time24 base classes.

第10行至第20行定义了一个从两个基类calendar和time24派生出的类 date_time。

第27行和第28行使用从基类calendar和time24继承的成员函数get_date( )与get_time( )获得dt的时间和日期值。

同理，第34行和第35行使用从基类calendar和time24继承的成员函数set_date( )与 set_time( )来设置dt的日期和时间值。

11.8　Virtual base classes（虚基类）

Consider the following class hierarchy (see Figure 11.6):

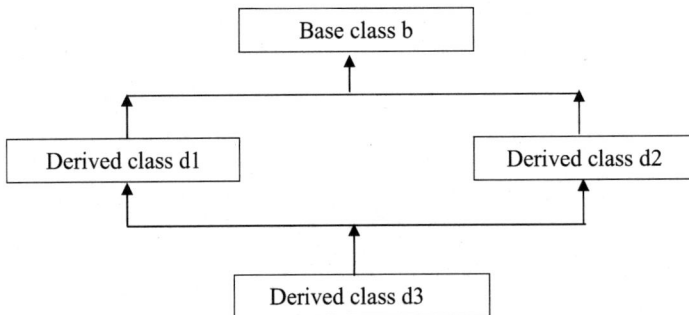

Figure 11.6　Multiple inheritance

Class b is the base class of both d1 and d2, which are the base classes of d3. The protected and public members of b will be inherited by d1 and d2 from their immediate base class b. Since

类b是d1和d2的基类，而类d1和d2又是d3的基类。这样类d1和d2就会从它们的直接基类b继承protected和public成员。因为d3

d3 is derived from both d1 and d2, it inherits the `protected` and `public` members of b from both d1 and d2. Two copies of the `protected` and `public` members of b now exist in d3. This is undesirable, as the next simple program shows.

是从d1和d2派生的，所以d3会从d1和d2中继承类b的protected和public成员。这样d3中就会有类b的protected和public成员的两份副本。这不是我们所期望的结果，正如下面的程序所显示的那样。

Program Example P11G

```
1  // Program Example P11G
2  // Demonstration of repeated inheritance.
3  #include <iostream>
4  using namespace std ;
5
6  class b
7  {
8  protected:
9    int b_data ;
10 public:
11   b()
12   {
13     cout << "Input a value for b_data " ;
14     cin >> b_data ;
15   }
16 } ;
17
18 class d1 : public b
19 {
20 protected:
21   int d1_data ;
22 public:
23   d1():d1_data( 0 )
24   {}
25 } ;
26
27 class d2 : public b
28 {
29 protected:
30   int d2_data ;
31 public:
32   d2():d2_data( 0 )
33   {}
34 } ;
35
36 class d3 : public d1, public d2
37 {
38 protected:
39   int d3_data ;
40 public:
41   d3():d3_data( 0 )
42   {}
43   void display_data() ;
44 } ;
```

```
45
46 void d3::display_data()
47 {
48   cout << "The two values of b_data are "
49        << d1::b_data << " and "
50        << d2::b_data << endl ;
51 }
52
53 int main( void )
54 {
55   d3 d3_obj ;
56   d3_obj.display_data() ;
57   return 0 ;
58 }
```

A sample run of this program is:

```
Input a value for b_data 1
Input a value for b_data 2
The two values of b_data are 1 and 2
```

Although artificial, the program demonstrates in a simple way the problem of *repeated inheritance*. When d3_obj is created on line 55, the base class constructor is called twice: once for class d1 and once for class d2. The data member b_data from the base class b is inherited twice in the derived class d3: once from d1 and once from d2. Lines 49 and 50 respectively display the value of b_data as inherited from d1 (d1::b_data) and the value of b_data as inherited from d2 (d2::b_data).

To avoid duplication of inherited members, the classes d1 and d2 must specify their base classes as a *virtual base class*. This is done in lines 18 and 27 by using the keyword virtual.

```
18 class d1 : virtual public b
27 class d2 : virtual public b
```

Only one set of protected and public members is now inherited in d3 from the base class b and the base class constructor is called only once for d3_obj.

Include the keyword virtual in lines 18 and 27 and re-run the program to see the result of using virtual base classes.

Now a sample run of this program is:

```
Input a value for b_data 1
The two values of b_data are 1 and 1
```

虽然上面程序的处理是我们人为设计的，但它以一种简单的方式说明了重复继承所带来的问题。在第55行创建对象d3_obj时，基类b的构造函数被调用了两次：一次是类d1调用的，另一次是类d2调用的。基类b的数据成员b_data也被派生类d3继承了两次：一次是从d1中继承的，另一次是从d2中继承的。第49行和第50行分别显示了从d1继承的b_data的值(d1::b_data)和从d2中继承的 b_data的值(d2::b_data)。

为了避免复制被继承的成员，类d1和d2必须将它们的基类指定为虚基类。

于是d3就只继承了一组基类b中的protected和public成员，并且在创建对象d3_obj时，基类的构造函数仅被调用一次。

Programming pitfalls

1. By default, the inheritance type is `private`, which is not the type that is usually required. So make sure to include the type of inheritance that is required, which is usually `public`.

2. A derived class automatically inherits all the `protected` and `public` data members of its base class, but not all of the member functions are automatically inherited. The following member functions are not inherited: constructors, destructors, overloaded operators and friends. The reason for this is that the code in constructors, destructors and overloaded operator functions are specific to a class and are unlikely to be of use in a derived class. Friends are not inherited because the friend of a base class need not necessarily be a friend of a derived class.

3. If a base class has no default constructor but has a constructor with a parameter list, then a derived class constructor must explicitly call the base class constructor in its initialisation list.

4. The keyword `virtual` must be used to avoid repeated inheritance of `protected` and `public` data members in inherited classes.

1. 默认情况下，继承类型为private（私有）继承，但这通常不是我们所需要的。因此应确认使用了需要的继承类型，通常为public（公有）继承。

2. 派生类自动继承基类所有的protected和public类型的数据成员，但并不自动继承所有的成员函数。下面这些基类的成员函数不会被派生类继承：构造函数、析构函数、重载运算符、友元。这是因为构造函数、析构函数、重载运算符函数是专门为一个类而设计的，在派生类中不太可能会用到它们。友元不能被继承是因为基类的友元没必要非要是派生类的友元。

3. 如果基类没有默认构造函数，但是有一个带有参数列表的构造函数，那么派生类构造函数必须在它的初始化列表中显式调用基类的构造函数。

4. 关键字virtual用来避免在派生类中重复继承protected和public类型的数据成员。

Quick syntax reference

	Syntax	Examples

Single Inheritance	`class derived : `*`access_level`*` base`
	`{`
	`// Data and functions.`
	`} ;`
	(*access_level* is either `public`,
	`protected` or `private`.)
		`class b`
		`{`
		`public:`
		`    ...`
		`protected:`
		`    ...`
		`private:`
		`    ...`
		`} ;`
		`class d : public b`
		`{`
		`public:`
		`    ...`
		`protected:`
		`    ...`
		`private:`
		`    ...`
		`} ;`

(cont.)

	Syntax	Examples

Multiple Inheritance	`class derived :` *access_level* `base1,`
	    access_level `base2,`
	    `. . .`
	`{`
	`// Data and functions.`
	`} ;`	

Exercises

1. (i) What type of relationship is involved in inheritance? Give an example.

   (ii) What type of relationship is involved in composition? Give an example.

2. In the code segment below, how many data members has the object `d_obj`?

```cpp
class b
{
protected:
  int x ;
  int y ;
private:
  int z ;
} ;
class d : public b
{
protected:
  float a ;
private:
  float b ;
} ;
int main( void )
{
  d d_obj ;
  return 0 ;
}
```

3. Explain the error in the following code segment:

```cpp
class b
{
private:
  float x ;
} ;
class d : public b
{
private:
  int y ;
public:
  void f()
  {
    y = x ;
```

```
    }
} ;
```
How is the error fixed?

4. Find and explain the errors in the following code segment:

```
class b
{
protected:
  float x, y ;
public:
  int z ;
} ;
class d : public b
{
public:
  void zero_x()
  {
    x = 0 ;
  }
} ;
int main( void )
{
  d d_obj ;
  d_obj.zero_x() ;
  y = x ;
  y = d_obj.x ;
  d_obj.y = d_obj.x ;
  return 0 ;
}
```

5. Explain what is wrong with this program segment?

```
class b
{
public:
  b( int v )
  {
    x = v ;
  }
private:
  int x ;
} ;
class d : public b
{
public:
  d( int v ) : y( v )
  {}
private:
  int y ;
} ;
int main( void )
```

```
{
  d d_obj( 1 ) ;
  return 0 ;
}
```

6. What is the output from the following program segment?

```
class b
{
public:
  b()
  {
    cout << "default constructor for b called" << endl ;
  }
} ;
class d1 : public b
{
public:
  d1()
  {
    cout << "default constructor for d1 called" << endl ;
  }
  d1( int v )
  {
  cout << "int parameter constructor for d1 called" << endl ;
  }
} ;
class d2 : public d1
{
public:
  d2()
  {
  cout << "default constructor for d2 called" << endl ;
  }
  d2( int v )
  {
  cout << "int parameter constructor for d2 called" << endl ;
  }
} ;

int main( void )
{
  b b_obj ;
  d1 d1_obj( 2 ) ;
  d1 d1_objs[ 2 ] ;
  d2 d2_obj( 3 ) ;
  return 0 ;
}
```

7. (i) How many copies of the data member m, n and o does an object of class d inherit?

```
class a
{
```

```
private:
  int m ;
  protected:
  int n ;
public:
  int o ;
  a()
  {
    cout << "default constructor for a called" << endl ;
  }
} ;
class b : public a
{
private:
  int p ;
} ;
class c : public a
{
private:
  int q ;
} ;
class d : public b, public c
{
private:
  int r ;
} ;
int main( void )
{
  class d d_obj ;
  return 0 ;
}
```

(ii) How many times is the constructor in the base class a called for the object d_obj?

8. Modify the code in the exercise 7 so that the output is the single line

```
default constructor for a called
```

9. Place the following two classes into a class hierarchy so that they inherit from a common base class:

```
class book                        class magazine
{                                 {
public:                           public:
  book() ;                          magazine() ;
private:                          private:
  string ISBN ;                     char frequency ;
  string title ;                    string title ;
  string author ;                   string publisher ;
  string publisher ;                string editor ;
  double price ;                    double price ;
} ;                               } ;
```

Write the base class with its constructor.

10. From the book class of exercise 9, use inheritance to derive a `library_book` class. This class will contain details of when a book was borrowed and by whom. For identification purposes, each library user has a six-digit library number.

11. Write a class, `postal_address`, to represent a postal address. The data members of this class are:

```
string addressline_1 ;
string addressline_2 ;
string addressline_3 ;
string addressline_4 ;
```

From this class derive a class called `inter_postal_address` for an international postal address. An international postal address has a country name in the address.

Write a default constructor for both the `postal_address` and `inter_postal_address`, which will initialise all the address lines to the empty string "".

Write a constructor for both classes to initialise the data members of the class with specified values. For example:

postal_address address1("HIT", "92 West Da-Zhi St.", "Harbin", "Heilongjiang") ;

inter_postal_address address2("DIT", "Kevin St.", "Dublin 8", "Leinster", "Ireland") ;

12. Using the class line from program P11E, write a program to construct two line objects and determine if the lines are at right angles to each other, i.e. the product of their slopes is −1.

Chapter Twelve
Polymorphism
第 12 章　多　　态

12.1　What is polymorphism?（什么是多态？）

The word *polymorphism* is derived from a Greek word meaning "many forms".

Polymorphism is one of the most important features in object-oriented programming and refers to the ability of different objects to respond differently to the same command (or 'message' in object-oriented programming terminology).

Polymorphism is used widely in our everyday language. For example, the command 'open' means different things when applied to different objects. Opening a bank account is very different from opening a window, which is different from opening a window on a computer screen. The command ('open') is the same, but depending on what it is applied to (the object) the resultant actions are different.

A very simple example of polymorphism is demonstrated in the next program.

"多态/多态性"这个词来源于希腊语，是"多种形式"的意思。

多态性是面向对象编程最重要的特征之一，是指不同的对象对同一命令（在面向对象编程术语中，命令又称为"消息"）做出不同响应的能力。

Program Example P12A

```
1   // Program Example P12A
2   // Simple demonstration of polymorphism.
3   #include <iostream>
4   using namespace std ;
5
6   class advanced_computer
7   {
8   public:
9     void hello( void )
10    {
11      cout << "Hello from the Advanced Computer" << endl ;
12    }
13  } ;
14
15  class simple_computer
16  {
17  public:
18    void hello( void )
19    {
```

```
20      cout << "Hello from the Simple Computer" << endl ;
21    }
22 } ;
23
24 int main( void )
25 {
26    advanced_computer HAL ;
27    simple_computer PC ;
28    HAL.hello() ;
29    PC.hello() ;
30    return 0 ;
31 }
```

Polymorphism:
The same command to different objects invokes different actions.

The output from this program is:

```
Hello from the Advanced Computer
Hello from the Simple Computer
```

It can be seen from the output that the objects HAL and PC have responded differently to the same hello() command sent to them on lines 28 and 29.

Polymorphism can be divided into two broad categories: *static* (or compile-time) and *dynamic* (or run-time) *polymorphism*. Static polymorphism occurs when the program is being compiled, dynamic polymorphism when the program is actually running.

C++ has three static polymorphism mechanisms: function overloading (chapter 7), operator overloading (chapter 10), and templates (chapter 13).

The use of the member function hello() in program P12A is an example of function overloading. In lines 28 and 29 the compiler inserts a function argument to identify the type of the calling object. The function headers on lines 9 and 18 are actually void hello (advanced_computer *this) and void hello (simple_computer *this), where this is a pointer to the object calling the function. Functions with the same names and different parameter lists mean function overloading.

When an operator is overloaded, a specific meaning is given to that operator. That meaning is dependent on the type of object that the operator is applied to. As seen in chapter 10, the meaning of the + operator, for example, can vary according to the type of object that it is applied to.

Before moving on to dynamic or run-time polymorphism, the next program shows an example of static polymorphism involving inheritance.

多态性主要分为两大类：静态（编译时）多态性和动态（运行时）多态性。静态多态性发生在程序被编译的时候，而动态多态性发生在程序运行的时候。

C++ 有 3 种静态多态性机制：函数重载（第 7 章），运算符重载（第 10 章），以及模板（第 13 章）。函数具有相同的名字及不同的参数列表就表示函数重载。

当运算符被重载时，这个运算符就被赋予了一个特定的意义，这种意义依赖于应用运算符的对象类型。

Program Example P12B

```
1  // Program Example P12B
2  // Demonstration of static polymorphism with inheritance.
3  #include <iostream>
4  using namespace std ;
5
6  const double pi = 3.14 ;
7
8  class circle
9  {
10 public:
11   circle( double r ) ;
12   double area( void ) ;
13 protected:
14   double radius ;
15 } ;
16 // circle member functions.
17 circle::circle( double r ) : radius( r )
18 {}
19
20 double circle::area( void )
21 {
22   return pi * radius * radius ;
23 }
24
25 // class cylinder derived from class circle.
26 class cylinder : public circle
27 {
28 public:
29   cylinder( double r, double l ) ;
30   double area( void ) ;
31 private:
32   double length ;
33 } ;
34 // cylinder member functions.
35 cylinder::cylinder( double r, double l ) : circle( r ), length( l )
36 {}
37
38 double cylinder::area( void )
39 {
40   return  2 * pi * radius * ( radius + length ) ;
41 }
42
43 // class sphere derived from class circle.
44 class sphere : public circle
45 {
46 public:
47   sphere( double r ) ;
48   double area( void ) ;
49 } ;
```

```
50 // sphere member functions.
51 sphere::sphere( double r ) : circle( r )
52 {}
53
54 double sphere::area( void )
55 {
56    return  4 * pi * radius * radius ;
57 }
58
59 int main( void )
60 {
61    circle circle_1( 1 ) ;
62    cylinder cylinder_1( 2, 3 ) ;
63    sphere sphere_1( 4 ) ;
64
65    cout << "Area of the circle: " << circle_1.area() << endl ;
66    cout << "Area of the cylinder: " << cylinder_1.area() << endl ;
67    cout << "Area of the sphere: " << sphere_1.area() << endl ;
68    return 0 ;
69 }
```

This program uses a simple class hierarchy in which a `cylinder` class and a `sphere` class are derived from a `circle` class. Each class has its own specific `area()` member function. Lines 65 to 67 are examples of polymorphism because the same command (`area()`) has been applied to the three different objects resulting in three different actions.

这个程序使用了一个简单的类层次结构，类cylinder和sphere都是从类circle派生的。每一个类都有自己特定的area( )成员函数。程序的第65行至第67行就说明了多态性，因为同一个命令［area( )］被应用于三个不同的对象，实现了三种不同的功能。

The output from this program is:

```
Area of the circle: 3.14
Area of the cylinder: 62.8
Area of the sphere: 200.96
```

12.2　Virtual functions（虚函数）

As demonstrated in the next program, polymorphism doesn't always work as intended.

As an example, let's modify the `circle` class of program P12B to include a new member function `area_message()` to display a message along with the value of the area.

Program Example P12C

```
1 // Program Example P12C
2 // Demonstration of polymorphism 'going wrong'.
3 #include <iostream>
4 #include <string>
5 using namespace std ;
6
7 const double pi = 3.14 ;
8
```

```
9  class circle
10 {
11 public:
12   circle( double r ) ;
13   double area( void ) ;
14   void area_message( string message ) ;
15 protected:
16   double radius ;
17 } ;
18 // circle member functions.
19 circle::circle( double r ) : radius( r )
20 {}
21
22 double circle::area( void )
23 {
24   return pi * radius * radius ;
25 }
26
27 void circle::area_message( string message )
28 {
29   cout << message << area() << endl ; Which area() is called?
30 }
31
32 // class cylinder derived from circle.
33 class cylinder : public circle
34 {
35 public:
36   cylinder( double r, double l ) ;
37   double area( void ) ;
38 private:
39   double length ;
40 } ;
41 // cylinder member functions.
42 cylinder::cylinder( double r, double l ) : circle( r ), length( l )
43 {}
44
45 double cylinder::area( void )
46 {
47   return 2 * pi * radius * ( radius + length ) ;
48 }
49
50 // class sphere derived from class circle.
51 class sphere : public circle
52 {
53 public:
54   sphere( double r ) ;
55   double area( void ) ;
56 } ;
57 // sphere member functions.
58 sphere::sphere( double r ) : circle( r )
```

```
59 {}
60
61 double sphere::area( void )
62 {
63   return  4 * pi * radius * radius ;
64 }
65
66 int main( void )
67 {
68   circle circle_1( 1 ) ;
69   cylinder cylinder_1( 2, 3 ) ;
70   sphere sphere_1( 4 ) ;
71
72   circle_1.area_message( "The area of the circle is: " ) ;
73   cylinder_1.area_message( "The area of the cylinder is: " ) ;
74   sphere_1.area_message( "The area of the sphere is: " ) ;
75   return 0 ;
76 }
```

The output from this program is:

```
The area of the circle is: 3.14
The area of the cylinder is: 12.56
The area of the sphere is: 50.24
```

The output from the program is not as expected and demonstrates the type of problem that can happen when using polymorphism.

What has happened is that the base class member function area() (lines 22 to 25) has been called (in line 29) for all objects, regardless of whether the object is a circle, a cylinder or a sphere. The derived class versions of area() (lines 45 to 48 and lines 61 to 64) should have been called for the derived objects cylinder_1 and sphere_1.

Normally, which function a call refers to is made at compile time. The compiler knows which function to call based on the object that calls it. This is called *static* or *early binding*.

通常，调用哪个函数是在编译时确定的。编译器根据调用函数的对象来确定调用哪个函数，这称为静态联编（也可译为静态绑定）或者早期联编。

Since the member function area_message() is in the base class, the compiler assumes that the call to area() on line 29 is a call to the base class version of area() - a reasonable assumption, but incorrect nevertheless.

In *dynamic* or *late binding*, it is left until run-time to determine which function should be called. This decision is based on the object making the call.

C++ uses virtual functions to implement dynamic binding. Virtual functions allow polymorphism to work in all situations.

The base class member function area() is made virtual by adding the keyword virtual to the function prototype on line 13.

在动态联编或者后期联编中，直到程序运行时才能确定调用哪个函数。即根据执行调用的对象确定调用哪个函数。

C++使用虚函数来实现动态联编。虚函数使得多态性在所有情况下都能起作用。

```
virtual double area() ;
```

Since the base class member function `area()` is virtual, the derived class versions of `area()` are also virtual. The keyword `virtual` is optional in the derived classes, although it is probably a good idea to include it for the sake of clarity.

Running the amended program, the output is now:

```
The area of the circle is: 3.14
The area of the cylinder is: 62.8
The area of the sphere is: 200.96
```

To further demonstrate dynamic binding, the next program asks the user to specify which shape to construct as the program is running. The compiler cannot know in advance which shape is going to be chosen. The program dynamically creates the shape during run-time and displays its area.

This example clearly shows that the virtual function `area()` and the object calling it are not bound together until run-time.

因为基类的成员函数area( )是虚函数，所以派生类的area( )也是虚函数。为了清楚起见，最好使用关键字virtual来声明虚函数，但是在派生类中，关键字virtual是可选的。

下面这个例子清晰地显示了虚函数area( )和调用它的对象是直到运行时才绑定到一起的。

Program Example P12D

```cpp
1  // Program Example P12D
2  // Demonstration of dynamic polymorphism.
3  #include <iostream>
4  #include <string>
5  using namespace std ;
6
7  const double pi = 3.14 ;
8
9  class circle
10 {
11 public:
12   circle( double r ) ;
13   virtual double area( void ) ;
14   void area_message( string message ) ;
15 protected:
16   double radius ;
17 } ;
18 // circle member functions.
19 circle::circle( double r ) : radius( r )
20 {}
21
22 double circle::area( void )
23 {
24   return pi * radius * radius ;
25 }
26
27 void circle::area_message( string message )
28 {
29   cout << message << area() << endl ;
30 }
```

```
31
32 // class cylinder derived from class circle.
33 class cylinder : public circle
34 {
35 public:
36   cylinder( double r, double l ) ;
37   virtual double area( void ) ;
38 private:
39   double length ;
40 } ;
41 // cylinder member functions.
42 cylinder::cylinder( double r, double l ) : circle( r ), length( l )
43 {}
44
45 double cylinder::area( void )
46 {
47   return  2 * pi * radius * ( radius + length ) ;
48 }
49
50 // class sphere derived from class circle.
51 class sphere : public circle
52 {
53 public:
54   sphere( double r ) ;
55   virtual double area( void ) ;
56 } ;
57 // sphere member functions.
58 sphere::sphere( double r ) : circle( r )
59 {}
60
61 double sphere::area( void )
62 {
63   return  4 * pi * radius * radius ;
64 }
65
66 int main( void )
67 {
68   char shape ;
69   double radius, height ;
70   circle* ptr ;
71
72   cout << "Enter a shape (1=circle, 2=cylinder 3=sphere) " ;
73   cin >> shape ;
74
75   if ( shape > '0' && shape < '4' )
76   {
77     cout << "Enter Radius " ;
78     cin >> radius ;
79
80     if ( shape == '1' )
```

```
81        {
82          ptr = new circle( radius ) ;
83        }
84        else if ( shape == '2' )
85        {
86          cout << "Enter Height " ;
87          cin >> height ;
88          ptr = new cylinder( radius, height ) ;
89        }
90        else if ( shape == '3' )
91        {
92          ptr = new sphere( radius ) ;
93        }
94        ptr -> area_message( "The area is: " ) ;
95      }
96      else
97        cout << "Invalid input" << endl ;
98      return 0 ;
99    }
```

A sample run of this program is:

```
Enter a shape (1=circle, 2=cylinder 3=sphere) 2
Enter Radius 3
Enter Height 4
The area is: 131.88
```

Line 70 defines `ptr` to be a pointer to the base class `circle`. Normally, a pointer to one type of data cannot be used to point to data of another type, e.g. a pointer to an `int` data type cannot point to a `float` data type. However, a pointer to a base class can also be used as a pointer to an object of a class derived from that base class. This is why `ptr` can be used in lines 82, 88 and 92 as a pointer to any of the newly created objects.

Line 94 uses `ptr` (which can be pointing to any of the objects) to call the member function `message_area()`.

12.2.1 When to use virtual functions

There are memory and execution time overheads associated with virtual functions, so be judicious in the choice of whether a base class function is virtual or not. In general, declare a base class member function as `virtual` if it may be overridden in a derived class.

12.2.2 Overriding and overloading

Both the terms *overriding* and *overloading* are used in object-oriented programming.

Overloading is a compiler technique that distinguishes between functions with the same name but with different parameter lists. Function overloading is covered in chapter 7.

程序的第70行将ptr定义为指向基类circle的指针。通常，指向一种数据类型的指针不能用来指向另一种数据类型。例如，一个指向int类型的指针不能用来指向float类型。然而，指向基类的指针却能用作指向该基类的派生类对象的指针。这就是为什么ptr能在程序的第82行、第88行和第92行用作指向任意新建对象指针的原因。第94行使用ptr（可以指向任意对象）来调用成员函数message_area( )。

使用虚函数会涉及内存和执行时间开销的问题，所以在决定是否将一个基类函数声明为虚函数时要慎重考虑。一般而言，如果一个基类的成员函数可能在派生类中被覆盖，那么就应将其声明为虚函数。

"覆盖"和"重载"都是面向对象编程的术语。

重载是一项编译器技术，用来区分函数名字相同但是参数列表不

Overriding occurs in inheritance when a member function of a derived class has the same name and the same parameters as a member function of its base class.

同的函数。函数重载已在第7章介绍。

覆盖发生在继承的时候，当派生类的成员函数与基类的成员函数的函数名和参数列表都相同时，便会发生覆盖。

12.3　Abstract base classes（抽象基类）

The next example program derives a deposit account class and a current account class using a bank account class as a base. The bank account class is similar to the one used in program P8G, but some changes have been made for the purposes of this example.

Program Example P12E

```
1    // Program Example P12E
2    // Demonstration of an abstract base class.
3    #include<iostream>
4    using namespace std ;
5
6    class bank_ac
7    {
8    public:
9      bank_ac() ;
10     void deposit( int amount ) ;
11     void withdraw( int amount ) ;
12   protected:
13     unsigned int ac_no ;
14     double balance ;
15   } ;
16   // bank_ac member functions.
17   bank_ac::bank_ac()
18   {
19     cout << "Enter account number: " ;
20     cin >> ac_no ;
21     balance = 0 ;
22   }
23
24   void bank_ac::deposit( int amount )
25   {
26     balance += amount ;
27   }
28
29   void bank_ac::withdraw( int amount )
30   {
31     balance -= amount ;
32   }
33
34   // deposit_ac class.
35   class deposit_ac : public bank_ac
36   {
37   public:
38     deposit_ac() ;
```

```
39    virtual void display_details( void ) ;
40  private:
41    double interest_rate ;
42  } ;
43  // deposit_ac member functions.
44  deposit_ac::deposit_ac()
45  {
46    cout << "Enter Interest Rate: " ;
47    cin >> interest_rate ;
48  }
49
50  void deposit_ac::display_details( void )
51  {
52    cout << "Deposit Account Details" << endl
53         << "Deposit Account No: " << ac_no << endl
54         << "Balance: " << balance << endl
55         << "Interest Rate: " << interest_rate << endl ;
56  }
57
58  // current_ac class.
59  class current_ac : public bank_ac
60  {
61  public:
62    current_ac() ;
63    virtual void display_details( void ) ;
64  private:
65    double overdraft_limit ;
66    double overdraft_rate ;
67    double standing_charges ;
68  } ;
69
70  // current_ac member functions.
71  current_ac::current_ac()
72  {
73    cout << "Enter Overdraft Limit: " ;
74    cin >> overdraft_limit ;
75    cout << "Enter Overdraft Rate: " ;
76    cin >> overdraft_rate ;
77    cout << "Enter Standing Charges: " ;
78    cin >> standing_charges ;
79  }
80
81  void current_ac::display_details( void )
82  {
83    cout << "Current Account Details" << endl
84         << "Current Account No: " << ac_no << endl
85         << "Balance: " << balance << endl
86         << "Overdraft Limit: " << overdraft_limit << endl
87         << "Overdraft Rate: " << overdraft_rate << endl
88         << "Standing Charges: " << standing_charges << endl ;
```

```
89  }
90
91  int main( void )
92  {
93    char menu( void ) ;
94    char choice ;
95    bank_ac *ptr ;
96
97    choice = menu() ;          // Get the type of account to create.
98    if ( choice == 'c' )       // Create a current_ac object.
99      ptr = new current_ac ;
100   else
101     ptr = new deposit_ac ;   // Create a deposit_ac object.
102
103   ptr -> display_details() ; // Display new account details.
104   return 0 ;
105 }
106
107 char mcnu( void )
108 // Purpose: Display a menu and return a choice.
109 {
110   char choice ;
111
112   cout << "New Account Set Up" << endl ;
113   do
114   {
115     cout << "Enter Account Type" << endl
116          << "C   Current A/C" << endl
117          << "D   Deposit A/C" << endl ;
118     cin >> choice ;
119     choice = tolower( choice ) ;
120   }
121   while( choice != 'd' && choice != 'c' ) ;
122
123   return choice ;
124 }
```

The program dynamically creates either a current_ac object on line 99 or a deposit_ac object on line 101. Note that although ptr is defined as a pointer to the base class object bank_ac, it is used as a pointer to objects of the derived classes current_ac and deposit_ac.

In line 103, ptr is used to display the details of the object it points to. The compiler doesn't know anything about dynamic binding and insists that since ptr is a pointer to the base class, then the base class should contain a member function display_details(), which it doesn't. The solution is to insert a 'dummy' display_details() member function in the public section of the base class:

这个程序要么动态创建一个 current_ac 对象（第99行），要么动态创建一个 deposit_ac 对象（第101行）。注意：虽然 ptr 被定义为一个指向基类对象 bank_ac 的指针，但它仍然可以用作指向其派生类 current_ac 和 deposit_ac 的对象。

程序第103行显示 ptr 所指向的对象的细节。在编译时不能进行动态联编，所以编译器认为既然 ptr 是指向基类的指针，那么基类中就应包含成员函数 display_details( )，但事实上它并不存在。为了解决这个问题，需要在基类的 public 部分插入一个虚拟的 display_details( )。

```
virtual void display_details() = 0 ;
```

The base class member function `display_details()` has no code and is assigned to 0.

Omitting this line from the public section of the base class `bank_ac` will result in a compiler error.

The text of the error will depend on the compiler but will state that the function `display_details()` is not a member of the base class `bank_ac`.

A class member function defined in this way is called a *pure virtual function*. A pure virtual function is required by the compiler but doesn't actually do anything.

A pure virtual function is overridden in all the derived classes and hence there is no need to implement it.

A sample run of program P12E is:

```
New Account Set Up
Enter Account Type
C Current A/C
D Deposit A/C
D
Enter account number: 12345
Enter Interest Rate: 5
Deposit Account Details
Deposit Account No: 12345
Balance: 0
Interest Rate: 5
```

A base class that contains a pure virtual function is called an *abstract base class*.

Abstract base classes (ABCs) are not used to create objects, but exist solely as a base for deriving other classes. It is not possible to define an object of an abstract base class, i.e. objects of an abstract base class cannot be instantiated.

Although a class hierarchy does not have to contain an abstract base class, typically one is defined during the first stage of developing a class hierarchy.

An abstract base class is usually the common denominator of more concrete derived classes (CDCs). The bank account class is a good example of an ABC. Go into a bank and ask to open a bank account and the first question asked will be "what type of account?". There is no such object as a bank account in the real world, but there are deposit and current account objects. A bank account is therefore an abstract concept. Other examples of ABCs are the building class and employee class used in chapter 8. Such classes are not useful on their own without the added details provided by their derived classes.

基类成员函数display_details( )内没有代码, 并且被赋值为0。

如果从基类bank_ac 的公有部分删掉这一行, 那么程序将产生一个编译错误。

错误提示信息依赖于所使用的编译器, 不过都会指出函数display_details( )不是基类bank_ac的成员函数。

用上面这种方式定义的类的成员函数称为纯虚函数。纯虚函数只是为了编译器的需要而存在的, 实际上并不做任何事情。

纯虚函数在所有的派生类中都被覆盖, 因此没有必要实现。

包含纯虚函数的基类称为抽象基类。

抽象基类（ABC）不是用来创建对象的, 它们的用途仅仅是作为基类去派生其他的类。定义一个抽象基类的对象是不可能的, 也就是说, 抽象基类的对象不能被实例化。

虽然一个类层次结构不必包含抽象基类, 但在形成类层次结构的第一阶段通常将定义一个抽象基类。

Programming pitfalls

1. It is easy to get confused between virtual member functions and virtual base classes. Although both are used in the context of inheritance, they are not in fact related. Perhaps a better name for virtual functions would be 'run-time-replaceable' functions?

2. An abstract base class must have at least one derived class. It is an error to attempt to define an object of an abstract base class.

3. A virtual member function in a base class can only be overridden in a derived class if the member function in the derived class has the same prototype as the base class member function.

4. Virtual functions in a base class can be overridden in one or more derived classes. A function must be declared as virtual in the base class; a derived class cannot make a function virtual. Unfortunately, this reduces the ability to reuse the base class without first modifying it.

5. The keyword `virtual` cannot be used outside the class declaration. It is a common error to include the keyword `virtual` when defining a non-inline member function.

```
class b
{
public:
  virtual void mf() ; // virtual is valid here.
} ;
virtual void b::mf() // error: virtual is not valid here.
{
  ...
}
```

1. 虚函数和虚基类很容易让人混淆。虽然它们都是用在继承的环境下，但是实际上它们并没有什么关系。

2. 抽象基类必须至少有一个派生类。定义一个抽象基类的对象将会引发错误。

3. 只有在派生类中的成员函数和基类的成员函数的函数原型相同时，基类中的这个虚函数才能被派生类中的函数覆盖。

4. 基类中的虚函数能在一个或者多个派生类中被覆盖。只能在基类中声明一个函数为虚函数，派生类中不能定义虚函数。遗憾的是，这将削弱重用基类的能力，要想重用基类，必须先对其进行修改。

5. 关键字virtual不能在类声明之外使用。在定义一个非内联成员函数时，使用关键字virtual是一个常见的错误。

Quick syntax reference

	Syntax	Examples
Declaring a virtual function	```class base { virtual member function } ;```	```class circle { public: virtual double area() ; ... } ;```
Declaring an abstract base class	```class base { virtual member function = 0 ; } ;```	```class bank_ac { public: virtual void display_details() = 0 ; ... } ;```

Exercises

1. What is the relationship between an abstract base class and a pure virtual function?
2. Explain the difference between late and early binding.
3. An abstract base class is a class containing a virtual function. True or false?
4. How can a base class force a derived class to include certain member functions?
5. Explain the error in the following program segment:

```
class b
{
public:
  virtual void mf( void ) ;
} ;
virtual void b::mf( void )
{
  cout << "member function mf called" << endl ;
}
```

6. What is the output from (a), (b), (c) and (d)?

   (a)
```
#include<iostream>
using namespace std ;

class b
{
public:
  void mf( int p ) ;
} ;
void b::mf( int p )
{
  cout << "member function mf in b called, "
       << "value of parameter is " << p << endl ;
}

class d : public b
{
public:
  void mf( int p ) ;
} ;
void d::mf( int p )
{
  cout << "member function mf in d called, "
       << "value of parameter is " << p << endl ;
}

int main( void )
{
  b* ptr ;
  d d_obj ;
  ptr = new d ;
  ptr -> mf( 1 ) ;
  d_obj.mf( 1 ) ;
}
```

(b)

```cpp
#include<iostream>
using namespace std ;

class b
{
public:
  virtual void mf( int p ) ;
} ;
void b::mf( int p )
{
  cout << "member function mf in b called, "
       << "value of parameter is " << p << endl ;
}

    class d : public b
{
public:
  void mf( int p ) ;
} ;
void d::mf( int p )
{
  cout << "member function mf in d called, "
       << "value of parameter is " << p << endl ;
}

int main( void )
{
  b* ptr ;
  d d_obj ;
  ptr = new d ;
  ptr -> mf( 1 ) ;
  d_obj.mf( 1 ) ;
  return 0 ;
}
```

(c)

```cpp
#include<iostream>
using namespace std ;

class b
{
public:
  void mf( int p ) ;
} ;
void b::mf( int p )
{
  cout << "member function mf in b called, "
       << "value of parameter is " << p << endl ;
}

class d : public b
{
```

```
public:
  void mf( double p ) ;
} ;
void d::mf( double p )
{
  cout << "member function mf in d called, "
       << "value of parameter is " << p << endl ;
}

int main( void )
{
  b *ptr ;
  d d_obj ;
  ptr = new d ;
  ptr -> mf( 1 ) ;
  d_obj.mf( 1.1 ) ;
  return 0 ;
}
```

(d)
```
#include<iostream>
using namespace std ;

class b
{
public:
  virtual void mf( int p ) ;
} ;
void b::mf( int p )
{
  cout << "member function mf in b called, "
       << "value of parameter is " << p << endl ;
}

class d : public b
{
public:
  void mf( double p ) ;
} ;
void d::mf( double p )
{
  cout << "member function mf in d called, "
       << "value of parameter is " << p << endl ;
}

int main( void )
{
  b *ptr ;
  d d_obj ;
  ptr = new d ;
  ptr -> mf( 1 ) ;
  d_obj.mf( 1.1 ) ;
  return 0 ;
}
```

Chapter Thirteen
Templates
第 13 章　模　　板

13.1　Introduction（引言）

A *template* is a framework for generating a class or a function. Instead of explicitly specifying the data types used in a class or a function, parameters are used. When arguments are assigned to the parameters, the class or function is generated by the compiler.

模板是生成类或函数的框架。与类或函数显式指定数据类型不同，模板使用形参。当把参数赋值给形参时，才由编译器生成类或函数。

13.2　Function templates（函数模板）

Consider the following function to find the maximum of two integer values.

```
int maximum( const int n1, const int n2 )
{
  if ( n1 > n2 )
    return n1 ;
  else
    return n2 ;
}
```

To find the maximum of two floating-point values a nearly identical function is required.

```
float maximum( const float n1, const float n2 )
{
  if ( n1 > n2 )
    return n1 ;
  else
    return n2 ;
}
```

The reason two different functions are required is simply that the function header is different in the two functions. The first function has parameters of type `int` and returns an `int`, the second function has parameters of type `float` and returns a `float`. The `if-else` statement in both functions are identical.

Function templates allow both of these functions to be replaced by a single function in which the parameter data types are replaced by a parameter `T` (for type) giving a generic or type-less function that can be used for all data types.

函数模板允许这样的两个函数被一个单独的函数取代，在这个单独的函数里，形参数据类型用T（T表示一种类型）表示，从而提供一个泛型或类型无关的函数，它适用于所有的数据类型。

Program Example P13A

```
1   // Program Example P13A
2   // Demonstration of a function template.
3   #include <iostream>
4   using namespace std ;
5
6   template <typename T>
7   T maximum( const T n1, const T n2 )
8   {
9     if ( n1 > n2 )
10      return n1 ;
11    else
12      return n2 ;
13  }
14
15  int main( void )
16  {
17    char c1 = 'a', c2 = 'b' ;
18    int i1 = 1, i2 = 2 ;
19    float f1 = 2.5, f2 = 3.5 ;
20
21    cout << maximum( c1, c2 ) << endl ; // a maximum() for chars
22    cout << maximum( i1, i2 ) << endl ; // a maximum() for ints
23    cout << maximum( f1, f2 ) << endl ; // a maximum() for floats
24    return 0 ;
25  }
```

T is called the type parameter. T is normally used, but any valid name can be used.

Running this program displays:

```
b
2
3.5
```

Line 6 is an example of a function template declaration. A function template declaration consists of the keyword `template` and a list of one or more data type parameters. The data type parameter is preceded by either the keyword `class` or the more meaningful keyword `typename`. Type parameters are enclosed within angle brackets < and >. When multiple data type parameters are used, they are separated by commas.

函数模板的声明由关键字template和包含一个或更多数据类型形参的参数列表构成。形参的数据类型前面可以是关键字class，也可以是意义更明确的关键字typename。类型形参T要用尖括号<和>括起来。当同时使用多种数据类型的形参时，它们之间要用逗号分开。

Line 21 has the effect of replacing the type parameter T in line 6 with the data type of the arguments c1 and c2, i.e. `char`. As a result, the compiler generates a function with the prototype

```
char maximum( const char n1, const char n2 ) ;
```

In line 22, the type parameter T is replaced by the data type of i1 and i2, i.e. `int`. This generates a function with the prototype

```
int maximum( const int n1, const int n2 ) ;
```

Similarly in line 23, T is replaced by the data type of f1 and f2, i.e.

`float`. This generates a function with the prototype

```
float maximum( const double, const double ) ;
```

This process is called *instantiation* of the template and the result is a conventional function generated by the compiler. (Technically, a function generated from a function template is known as a template function.)

Instead of using three separate functions in the program, a single function template is used. The program source file is smaller, but since the compiler generates three separate functions from the template, the executable file size remains the same.

Template definitions can be placed in a header file. Placing lines 6 to 13 of program P13A into a header file `maximum.h`, this file can be `#include`d into a program file. This is done in the next program.

这个过程称为模板的实例化，其结果是由编译器产生的一个常规函数。（严格地讲，用模板生成的函数称为模板函数。）

在上面的程序中，没有使用3个独立的函数，取而代之的是一个单独的函数模板，使得程序的源文件变小，但是因为编译器利用模板生成了3个不同的函数，所以可执行文件的大小仍然是不变的。模板的定义可以放在头文件里。例如，下面的程序就是把程序P13A的第6行至第13行的模板定义放到了头文件maximum.h里，然后再用#include将这个头文件包含到程序文件中。

Program Example P13B

```
1   // Program Example P13B
2   // Demonstration of including a function template in a program.
3   #include<iostream>
4   #include "maximum.h"
5   using namespace std ;
6
7   int main( void )
8   {
9       char c1 = 'a', c2 = 'b' ;
10      int i1 = 1, i2 = 2 ;
11      float f1 = 2.5, f2 = 3.5 ;
12
13      cout << maximum( c1, c2 ) << endl ;
14      cout << maximum( i1, i2 ) << endl ;
15      cout << maximum( f1, f2 ) << endl ;
16      return 0 ;
17  }
```

Function templates are very useful when a program requires a function to sort the elements of an array. If the sorting algorithm is implemented as a function template, functions to sort all types of arrays can be automatically generated by the compiler.

The next program uses the bubble sort algorithm to sort and display arrays of different data types. The program uses three function templates contained in the header files `bubble.h`, `swap.h` and `show.h`.

当程序需要对数组元素进行排序时，函数模板是非常有用的。如果排序算法能用函数模板来实现，那么编译器就能自动生成对任何类型的数组都适用的排序函数。

Program Example P13C

```
1   // Program Example P13C
2   // Demonstration of three function templates to sort
3   // and display arrays of any data type.
4   #include <iostream>
5   #include "swap.h"
6   #include "bubble.h"
7   #include "show_array.h"
```

Header files containing function templates.

```
8  using namespace std ;
9
10 int main( void )
11 {
12   int int_array[] = { 23, 4, 12, -9, 55 } ;
13   float float_array[] = { 4.2, 7.8, 77.3, 1.2 } ;
14
15   bubble_sort( int_array, 5 ) ;
16   show_array( int_array, 5 ) ;
17   bubble_sort( float_array, 4 ) ;
18   show_array( float_array, 4 ) ;
19   return 0 ;
20 }
```

Use of function templates with different data types.

Contents of bubble.h:

```
1  // Declaration of the bubble function template.
2  #if !defined BUBBLE_T_H
3  #define BUBBLE_T_H
4
5  template <typename T>
6  void bubble_sort( T array[], int n )
7  // Purpose   : Sorts an array using the bubble sort algorithm.
8  // Parameters: The array to sort and
9  //             The number of elements in the array.
10 {
11   bool swapping = true ;
12   while( swapping )
13   {
14     swapping = false ;
15     for ( int i = 0 ; i < n-1 ; i++ )
16     {
17       if ( array[i] > array[i+1] )
18       {
19         swapping = true ;
20         swap_values( array[i], array[i+1] ) ;
21       }
22     }
23   }
24 }
25
26 #endif
```

Contents of swap.h:

```
1  // Declaration of the swap function template.
2  #if !defined SWAP_T_H
3  #define SWAP_T_H
4
5  template <typename T>
6  void swap_values( T &x, T &y )
7  // Purpose   : Swap two values.
8  // Parameters: Two values of any data type.
```

```
9  {
10   T temp = x ;
11   x = y ;
12   y = temp ;
13 }
14
15 #endif
```

Contents of `show_array.h`:

```
1  // Declaration of the show_array function template.
2  #if !defined SHOW_ARRAY_T_H
3  #define SHOW_ARRAY_T_H
4
5  using namespace std ;
6  template <typename T>
7  void show_array( T array[], int n )
8  // Purpose   : Displays the elements of an array.
9  // Parameters: array[] - the array to display.
10 //             n - the number of elements in the array.
11
12 {
13   for ( int i = 0 ; i < n ; i++ )
14     cout << array[i] << ' ' ;
15   cout << endl ;
16 }
17
18 #endif
```

As with previous uses of header files, the preprocessor statements are used in each of these header files to prevent the header file from being inadvertently included more than once in a program.

正如前面使用头文件那样，需要使用编译预处理命令来避免这些头文件被多次重复包含于同一个程序中。

13.3　Class templates（类模板）

As well as generating functions, templates can be also used for generating entire C++ classes.

模板不仅能生成函数，还能用来生成完整的C++类。

Common data structures such as linked lists and stacks are independent of the data type they contain. Instead of having a class for a stack of integers, and another class for a stack of floating-point values, and yet another class for a stack of strings, templates allow one class to be written for all data types.

通用数据结构，例如链表和堆栈，它们都与包含的数据类型无关。模板允许一个类用于所有数据类型，不必再为整型、浮点型或者字符型分别写一个堆栈类。

Class templates are also known as *parameterised types*. A parameterised type is a data type defined in terms of other data types, some of which are unspecified.

类模板也称为参数化类型。参数化类型是由其他数据类型甚至是一些未指定的数据类型来定义的数据类型。

To demonstrate the use of class templates, the next example uses a class template for a stack data structure.

The operations that can be performed on a stack of plates, whereby plates are only added and removed from the top of the stack, are

可以用"对一摞盘子的操作"来形容对一个堆栈数据结构的操

analogous to the operations that can be performed on a stack data structure.

A stack data structure is implemented with two principal operations:

push: adds an item to the top of the stack.

pop : retrieves the most recently pushed item from the stack.

For example,

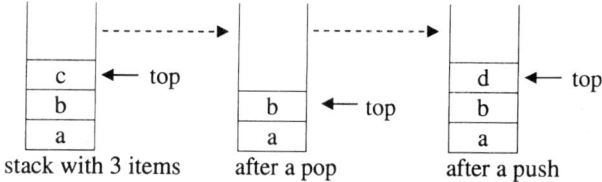

stack with 3 items　　after a pop　　after a push

A simple way to implement a stack is to use an array to hold the items and a variable to hold the index value of the 'top' of the array.

A stack can therefore be defined as a class template with a type parameter that specifies the data type of the items to be stored on the stack and an integer variable to indicate the current top of the stack.

The definition of a stack class template is stored in a header file stack.h containing the following:

```
1   // Declaration of stack class template.
2   #if !defined STACK_T_H
3   #define STACK_T_H
4
5   template <typename T>
6   class stack
7   {
8   public:
9     stack( int n = 10 ) ;
10    // Purpose   : Class constructor.
11    // Parameters: n - size of stack.
12    ~stack() ;
13    // Purpose   : Class destructor.
14    bool pop( T& data_item ) ;
15    // Purpose   : Remove an item from the stack.
16    // Parameters: data_item - data item on top of the stack is
17    //             copied to this reference.
18    // Returns   : true - operation successful.
19    //             false - operation failed.
20    bool push( const T& data_item ) ;
21    // Purpose   : Add an item to the stack.
22    // Parameters: data_item is placed on top of the stack.
23    // Returns   : true - operation successful.
24    //             false - operation failed.
25    inline int number_stacked() const ;
26    // Purpose   : Inspector function.
```

作，放盘子和取盘子只能在这一摞盘子的顶部进行。

一个堆栈数据结构包含两种主要的操作：

压栈：在堆栈顶端添加一项。

出栈：从堆栈顶端弹出最近一次压入堆栈中的项。

实现堆栈的一种简单的方法是使用一个数组保存压入堆栈中的项，使用一个变量记录数组"顶端"的下标值。

因此，堆栈可以定义为类模板，使用类型形参来指定要存储的数据项的类型，使用一个整型变量来指示当前的栈顶。

```
27   // Returns  : Number of items currently on the stack.
28    inline int stack_size( void ) const ;
29   // Purpose  : Inspector function.
30   // Returns  : Size of the stack.
31 private:
32   int max_size ;
33   int top ;
34   T* data ;
35 } ;
36 // stack member functions
37 template <typename T>
38 stack <T>::stack( int n )
39 {
40   max_size = n ;
41   top = -1 ;
42   data = new T[ n ] ;
43 }
44
45 template <typename T>
46 stack <T>::~stack()
47 {
48   delete [] data ;
49 }
50
51 template <typename T>
52 bool stack <T>::push( const T& data_item )
53 {
54   if (  top  <  max_size -1 )
55   {
56     data[ ++top ] = data_item ;
57     return true ;
58   }
59   else
60     return false ;
61 }
62
63 template <typename T>
64 bool stack <T>::pop( T& data_item )
65 {
66   if ( top > -1 )
67   {
68     data_item = data [ top-- ] ;
69     return true ;
70   }
71   else
72     return false ;
73 }
74
75 template <typename T>
76 int stack <T>::number_stacked() const
```

```
77 {
78   return top + 1 ;
79 }
80
81 template <typename T>
82 int stack <T>::stack_size( void ) const
83 {
84   return max_size ;
85 }
86
87 #endif
```

A reference to a class template must include its parameter list. This is why the definition of the member functions use stack <T> instead of just stack, which would be used if the class were not a template. The next program demonstrates the use of the stack class template.

引用类模板时必须包含它的形参列表。这就是成员函数的定义使用stack <T>而不是仅使用stack的原因，如果堆栈类不是模板，则仅使用stack。

Program Example P13D

```
1   // Program Example P13D
2   // Demonstration of a class template.
3   #include <iostream>
4   #include "stack.h"
5   using namespace std ;
6
7   int main( void )
8   {
9     stack <int> i_stack ;        // A stack to hold 10 integers.
10    stack <char> c_stack( 5 ) ; // A stack to hold 5 characters.
11    int i, n, int_data ;
12    char char_data ;
13    bool success ;
14
15    // Push some values onto c_stack.
16    c_stack.push( 'a' ) ;
17    c_stack.push( 'b' ) ;
18    c_stack.push( 'c' ) ;
19
20    // Use a for loop to fill i_stack.
21    n = i_stack.stack_size() ;
22    for( i = 0 ; i < n ; i++ )
23      i_stack.push( i ) ;
24
25    // Display the values on c_stack.
26    cout << "character stack data:" << endl ;
27    n = c_stack.number_stacked() ;
28    for ( i = 0 ; i < n ; i++ )
29    {
30      success = c_stack.pop( char_data ) ;
31      if ( success )
32         cout << char_data << ' ' ;
```

```
33    }
34    cout << endl ;
35
36    // Display the values on i_stack.
37    cout << "integer stack data:" << endl ;
38    n = i_stack.number_stacked() ;
39    for ( i = 0 ; i < n ; i++ )
40    {
41      success = i_stack.pop ( int_data ) ;
42      if ( success )
43        cout << int_data << ' ' ;
44    }
45    cout << endl ;
46    return 0 ;
47 }
```

The data type of the stack data is specified between the angle brackets < and >.

Line 9 generates a stack class from the class template to process up to ten integer data items (ten is the default value in the class constructor). Line 10 generates a stack class from the class template to process up to five character data items. (Technically, a class generated from a class template is known as a template class.)

Lines 16 to 18 and lines 22 to 23 push some sample data items onto each of the stack objects c_stack and i_stack. Lines 28 to 33 and lines 39 to 44 pop and display data from each of the stacks.

The output from program P13D is:

```
character stack data:
c b a
integer stack data:
9 8 7 6 5 4 3 2 1 0
```

Compared with non-template functions and classes, function templates and class templates look daunting. To create templates, first write a non-template data type specific version of the function or class. When the data type specific version is working satisfactorily, replace the specific data types with template parameters. For example, the bubble sort function template was initially developed with an int data type. When the function was working for integer data, the data type int was replaced with the type parameter T, enabling it to work with any data type.

堆栈数据的类型由两个尖括号< 和>之间的数据类型指定。

第9行用类模板生成了一个堆栈 类，以处理10个整型数据项（10是 类的构造函数的默认值）。

第10行用类模板生成了一个堆栈 类，以处理5个字符数据项。（严 格地讲，从类模板生成的类称为 模板类。）

与非模板函数和非模板类相比， 函数模板和类模板看上去更"令 人生畏"。为了创建模板，首先要 编写一个非模板特定数据类型的 函数或者类。当这个指定数据类 型的函数或者类运行得令人满意 之后，再用模板形参来取代其中 的特定数据类型。例如，设计冒 泡排序函数模板时，先使用int数 据类型来设计。当这个函数对于 整型数据很好用之后，就可以用 类型形参T来替代int，从而使得 这个冒泡排序对所有的数据类型 都适用。

Programming pitfalls

1. Do not assume that all operators in a function prototype are defined for the data type passed to the template parameter. For example, using the function template `maximum()` in program P13A for objects of a point class,

```
point p( 1, 2 ), p2( 3, 4 ), p3( 0, 0 ) ;
p3 = maximum( p1, p2 ) ;
```

An error occurs if > is not overloaded in the definition of the point class.

1. 不要以为函数原型中的所有运算符都适合作为传递给模板形参的数据类型。

2. The keyword `typename` must be included for each template parameter.

For example,

```
template <typename T1, T2>
```

will give an error because `typename` should also precede `T2`.

2. 每一个模板形参前都必须写上关键字typename。

3. Each specified type parameter must be used in the function template. For example,

```
template <typename T1, typename T2>
int my_function ( T1 var ) // Error - T2 is not used!
```

will give an error because `T2` is not used.

3. 每一个指定的类型形参都必须在函数中使用。

Quick syntax reference

	Syntax	Examples

Defining a function template	`template <typename T1,`
	`            typename T2, ...>`
	`// Function definition.`
		`template <typename T>`
		`T maximum( const T n1,`
		`            const T n2 )`
		`{`
		`  ...`
		`}`

Defining a class template	`template <typename T1,`
	`            typename T2, ...>`
	`// Class definition.`
		`template <typename T>`
		`class stack`
		`{`
		`  ...`
		`} ;`

Exercises

1. Write a function template `largest_element()` that returns the largest value in an array. The array may contain elements of any one data type. The function has two parameters. The first parameter is the name of the array and the second parameter is the integer number of elements in the array. The return type of the function is the same type as its first parameter.

2. Write a function template `locate()` that finds and returns the first position of a value in an array. The array may contain elements of any one data type. The function has two parameters. The first parameter

is the name of the array and the second parameter is a variable of the same type as the first parameter. If the value does not exist in the array, -1 is returned.

3. Write a function template `power()` that returns the value of its first parameter raised to the power of its second parameter. The first parameter may have any data type and the second parameter is an integer. The return type of the function is the same type as its first parameter.

4. Convert the following function into a function template.

```cpp
void sort_ascend ( int& a, int &b )
{
  if ( a > b )
  {
    int t = a ;
    a = b ;
    b = a ;
  }
}
```

5. The binary search or binary chop algorithm is used to search for a value in a sorted array. The algorithm is analogous to searching for a name in a telephone book or looking up the meaning of a word in a dictionary.

The following function searches a sorted array a of n integer elements for an integer value held in `item`.

```cpp
int bin_search( int a[], int n, int item )
{
  int mid ;
  int top = n - 1 ;
  int bottom = 0 ;

  while( top >= bottom )
  {
    mid = ( top + bottom ) / 2 ;

    if ( item == a[ mid ] ) // Is the item found?
      return mid ;          // Yes - return the index value.

    if ( item > a[mid] )    // No - the item is not found.
      bottom = mid + 1 ;    // Try the top half.
    else
      top = mid - 1 ;       // Try the bottom half.
  }
  return -1 ;               // The item is not found.
}
```

Convert the function to a function template so that it will work with any data type.

Test the function template by instantiating `bin_search()` with T bound to a built-in type such as `double`. For example,

```cpp
double numbers[] = { 0.8, 1.1, 1.2, 3.9, 8.7, 15.92, 71.9, 80.31} ;
cout << bin_search( numbers, 8, 1.2 ) << endl ;
```

6. Add a new member function `is_full()` to the stack class template. This member function returns `true` or `false`, depending on whether the stack contains its maximum number of values or not.

7. The member function `push()` of the stack class template simply returns `false` if the stack is full. Modify `push()` so that the stack size is dynamically increased by 10 to allow for ten more data items. The data items already on the stack should not be lost.

8. Write a class template for a Queue class. A Queue is like a stack, except that insertions are at the end and removals are from the front. It simulates an everyday waiting queue.

   Test the template with the following program:

```
int main( void )
{
  queue<char> q( 3 ) ;
  q.insert( 'A' ) ;      // Like push(), except at the end.
  q.insert( 'B' ) ;
  q.insert( 'C' ) ;
  cout << q.remove() ; // Same as pop(), except from the front.
                       // Should display A.
  cout << q.remove() ; // Should display B.
  cout << q.remove() ; // Should display C.
  return 0 ;
}
```

The output from this program should be:

```
ABC
```

Chapter Fourteen
Files and Streams
第 14 章　文 件 和 流

The input-output statements used so far were those that read data from the keyboard and displayed data on the screen. When a program reads data from the keyboard, it stores the data in the computer's memory. When the program terminates, the data is lost and must be reentered every time the program is run. This chapter covers file stream input and output. Unlike data that is kept in the computer's memory, files use external storage devices, such as hard disks and USB keys to store data. These are permanent storage devices that allow data to be stored after the program terminates.

到目前为止，我们所用的输入输出语句都是从键盘读取数据的，然后在屏幕上显示出来。当程序从键盘读取数据的时候，它把数据存储在计算机的内存中。当程序终止的时候，内存中的数据就会丢失。这样，每次运行程序的时候都要重新输入这些数据。本章介绍文件输入/输出流。与存储在计算机内存中的数据不同的是，文件使用外部存储设备（例如硬盘和U盘）来存储数据。它们都是永久性的存储设备，保存的数据在程序终止时不会丢失。

14.1　The C++ input/output class hierarchy
（C++ 输入/ 输出类的层次结构）

Unlike many programming languages, C++ has no built-in input/output (I/O) commands. In C++ the I/O commands are included in a class library. A simplified diagram of the I/O class hierarchy is shown in Figure 14.1 and a brief explanation of the hierarchy follows.

与其他许多编程语言不同，C++没有内置的输入/输出（I/O）命令。在C++中，I/O命令被包含在类库中。

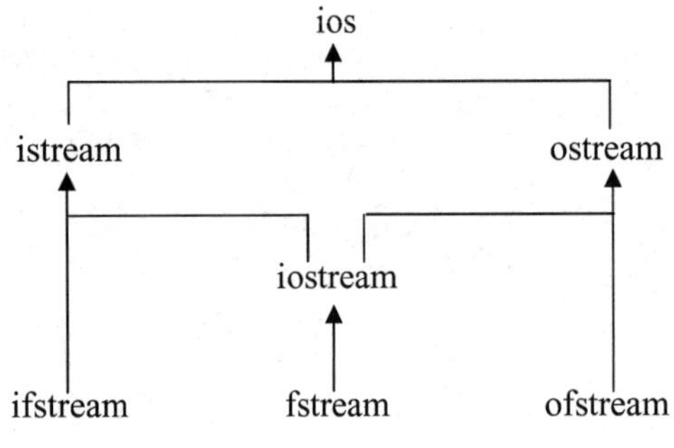

Figure 14.1　I-O Class hierarchy

The base class `ios` contains data members to indicate conditions such as whether a stream object is opened for input or output and whether the end of the file has been reached.

The derived class `istream` extends the base class by adding member functions to input data from a stream. This class provides basic input processing by including overloaded operators >> for reading the built-in data types (`char`, `int`, `float` etc.) from a stream. As seen

基类ios中有一些数据成员是用来描述状态的，例如输入/输出流对象是否处于打开状态，以及是否已经到达文件的末尾等。

派生类istream向基类添加了一些从流中读取数据的函数。它通过重载运算符>>来从流中读取内置数据类型（char，int，float，等

in program P10G, additional >> overloads can be written for user-defined classes. The stream `cin`, which is normally attached to the keyboard, is an object of this class.

The derived class `ostream` contains member functions to output data to a stream. This class provides basic output processing by including overloaded insertion operators (<<) for writing built-in data types to a stream. Additional overloaded insertion operators can be written for user-defined classes (see program P10G). The stream `cout` is an object of this class and is normally attached to the screen.

The class `ifstream` is derived from `istream` and is used to create input file objects.

The class `ofstream` is derived from `ostream` and is used to create output file objects.

Finally, the class `fstream` is used to create file objects that can be used for both input and output.

The definition of the library classes is contained in the header files `iostream` and `fstream`, summarised in Table 14.1.

等）的数据，从而提供基本的输入功能。正如程序P10G所示，可以为用户自定义的类添加>>重载函数。流cin是派生类istream的对象，通常和键盘输入相关联。

派生类ostream中包含的一些成员函数可用于向流中输出数据。它通过重载运算符<<来向流中输出内置数据类型的数据，从而提供了基本的输出功能。同理，程序P10G也为用户自定义的类添加了<<重载函数。流cout是派生类ostream的对象，通常和屏幕输出相关联。

ifstream是从istream派生的类，用于创建输入文件对象。

ofstream则是从ostream派生的类，用于创建输出文件对象。

而用于创建文件对象的类fstream既能用于输入，也能用于输出。

Table 14.1   Summary of `iostream` and `fstream` header files

`iostream`	Includes the definition of
	• `ios`, `istream`, `ostream`, and `iostream` classes
	• stream objects `cin`, `cout`, `clog` and `cerr`
	• stream manipulators `endl`, `ws`, `dec`, `hex` and `oct`.
	• Includes the definition of the `ifstream`, `ofstream`,
`fstream`	and `fstream` classes.

14.2　Opening a file（打开文件）

To open a file, first create an instance of the appropriate class, as shown in the following statements:

为了打开一个文件，首先要创建一个适当的类的实例。

```
ifstrem in ;      // in is an input file object
ofstream out ;    // out is an output file object
```

After creating an instance of the appropriate class, the file object must be opened to associate it with a file stored on a hard disk or some other storage device.

在创建一个适当的类的实例之后，必须打开该文件对象，将其和存储在硬盘或其他存储设备上的文件相关联。

```
in.open( "in.dat" ) ;    // in is associated with in.dat
out.open( "out.dat" ) ;  // out is associated with out.dat
```

Creating an instance of a file object and opening a file can be combined into one statement by using the `ifstream` and `ofstream` class constructors.

创建文件对象的实例和打开文件可以合并为一条语句来完成，这可以通过使用类ifstream和ofstream的构造函数来实现。

```
ifstrem in( "in.dat" ) ;
ofstream out( "out.dat" ) ;
```

It is good practice to close a file after processing it.

在文件处理完毕后立即关闭该文件，是一个良好的编程习惯。

```
in.close() ;   // Close input stream in, i.e. in.dat
out.close() ; // Close output stream out, i.e. out.dat
```

The next program is a simple demonstration of writing data to a file and reading the same data back again. Both methods of opening a file are demonstrated.

Program Example P14A

```cpp
1    // Program Example P14A
2    // Demonstration of file output and file input.
3    #include <iostream>
4    #include <fstream>
5    #include <string>
6    using namespace std ;
7
8    int main( void )
9    {
10     char c = 'A' ;
11     int i = 1 ;
12     string s = "Hello" ;
13
14     ofstream out ;            // Create a file object out and
15     out.open( "file.txt" ) ;  // open file.txt
16
17     // Write some data to the file using <<
18     out << s << ' ' << c << ' ' << i << endl ;
19     out.close() ; // Close the output file.
20
21     ifstream in( "file.txt" ) ;  // Use constructor to open the file.
22     // Read data from the file using >>
23     in >> s >> c >> i ;
24     in.close() ; // Close the input file.
25
26     cout << "Data read from file:" ;
27     cout << s << ' ' << c << ' ' << i << endl ;
28     return 0 ;
29   }
```

The output from this program is:

```
Data read from file:Hello A 1
```

Note that the header file `fstream` is included on line 4.

Line 14 creates an instance `out` of an output file object. Line 15 associates `out` with the external file `file.txt` and opens the file for processing. If the file does not exist, by default it will be created in the same folder (directory) as the program.

Line 18 outputs some data to the file stream using << in the same way that data would be sent to the stream `cout`.

The `ifstream` class constructor is used on line 21 to open `file. txt` for input processing.

第14行创建了一个输出文件对象的实例out。第15行把out与一个外部文件file.txt相关联，并且打开这个文件用于处理操作。如果该文件不存在，那么默认情况下将在程序所在的同一文件夹（目录）下创建这个文件。

第18行按照将数据输出到流对象cout中的同样方式，使用<<将数据输出到文件流中。

第21行使用ifstream类的构造函数来打开file.txt文件，准备进行输入操作。

Line 23 is similar to the way data is input from the keyboard, except `in` is used instead of `cin`.

The contents of `file.txt` can be viewed using any text editor such as Windows Notepad or Linux gedit.

14.3　File error checking（文件出错检查）

It is good practice to check for errors when opening a file. Possible errors that can occur include errors due to no space being available on an output device or attempting to open a non-existent file for reading. The `ios` class contains member functions that can be used to check for errors. Since `ifstream` and `ofstream` are inherited from `ios`, these functions are available for use with `ifstream` and `ofstream` objects. Some of the member functions along with their return values are given in Table 14.2.

Table 14.2　Some member functions of `ios`

Member Function	Return Value
`fail()`	Returns the Boolean value `true` if the operation failed; otherwise the Boolean value `false` is returned.
`good()`	Returns the Boolean value `true` if the operation succeeded; otherwise the Boolean value `false` is returned. This is the opposite of `fail()`.
`eof()`	Returns the Boolean value `true` if the end of the file has been reached; otherwise the Boolean value `false` is returned.

The next program modifies program P14A by using the member function `fail()` to check for errors when opening a file. If an error does occur, the program returns 1 to the operating system to indicate that an error condition has arisen; otherwise 0 is returned to the operating system on line 36.

(Note: by convention, a return value of zero indicates success, while a non-zero return value indicates an error.)

Program Example P14B

```
1   // Program Example P14B
2   // Program to demonstrate file open error checking.
3   #include <iostream>
4   #include <fstream>
5   #include <string>
6   using namespace std ;
7
8   int main( void )
9   {
10    char c = 'A' ;
11    int i = 1 ;
12    string s = "Hello" ;
```

第23行读取文件的方法和从键盘读入数据的方法类似，唯一不同的就是用in代替了cin。

在打开文件时，检查是否发生错误，是一个良好的编程习惯。文件打开时可能发生的错误包括输出设备上没有可用的空间，或者要读取的文件并不存在。

ios类中包含的一些成员函数可用来检查这类错误。因为ifstream和ofstream继承自ios类，所以这两个类的对象也能使用这些成员函数。表14.2列出了一些成员函数和它们的返回值。

下面的程序修改了程序P14A，使用成员函数fail( )来检查打开文件时可能发生的错误。如果的确发生了错误，那么程序向操作系统返回1，用1表示出现了一个错误的状态。否则如第36行所示，向操作系统返回0。

（注意：按照惯例，返回零值表示成功，返回非零值表示出现了一个错误。）

```
13   ofstream out ;
14
15   out.open ( "file.txt" ) ;
16   if ( out.fail() ) // Has the file failed to open?
17   {
18     cerr << "Failure to open file.txt for output" << endl ;
19     return 1 ;
20   }
21   // Write some data to the file using <<
22   out << s << ' ' << c << ' ' << i << endl ;
23   out.close() ;
24
25   ifstream in( "file.txt" ) ;
26   if ( in.fail() )
27   {
28     cerr << "Failure to open file.txt for input" << endl ;
29     return 1 ;
30   }
31   // Read data from the file using >>
32   in >> s >> c >> i ;
33   in.close() ; // Close the input file.
34   cout << "Data read from file:" ;
35   cout << s << ' ' << c << ' ' << i << endl ;
36   return 0 ;
37 }
```

The additional program statements can be tested by adding a non-existent storage device identifier to the file path, e.g. instead of "file.txt" specify "xx:/file.txt" on lines 15 and 25.

14.4   Single character I/O and detecting the end of a file（单字符的 I/O 和文件末尾的检测）

When reading data from an input file, it is necessary to be able to detect the end of the file so that processing of the file may stop. The end of a file may be detected using the ios member function eof() inherited by the ifstream class.

The next program demonstrates the use of eof() in making a copy of a text file. The istream member function get() is used to read a character from the input file and the ofstream member function put() is used to write the character to an output file.

从文件中读取数据时，必须检测是否到达了文件的末尾，以便结束对文件的处理操作。可以使用 ios 类的成员函数 eof( ) 来检测文件的末尾，ifstream 类也从 ios 类继承了这个函数。

下面的程序演示了在复制文本文件时函数 eof( ) 的作用。类 istream 的成员函数 get( ) 用于从输入文件中读取一个字符，类 ofstream 的成员函数 put( ) 则负责将字符写入输出文件中。

Program Example P14C

```
1  // Program Example P14C
2  // Program to demonstrate member functions eof, get and put.
3  #include <iostream>
4  #include <fstream>
5  using namespace std ;
6
```

```
7   int main( void )
8   {
9     char c ;
10    int count ;
11
12    ifstream in( "file.txt" ) ;   // Open the input file.
13    if ( in.fail() )
14    {
15      cerr << "Open failure on file.txt" << endl ;
16      return 1 ;
17    }
18
19    ofstream out( "file.bak" ) ;   // Open output file.
20    if ( out.fail() )
21    {
22      cerr << "Open failure on file.bak" << endl ;
23      return 1 ;
24    }
25
26    in.get( c ) ; // Read the first character in the file.
27    count = 1 ;
28    while( !in.eof() ) // Loop while not end of file.
29    {
30      out.put( c ) ;   // Write the character.
31      in.get( c ) ;    // Read the next character.
32      count++ ;
33    }
34
35    in.close() ;
36    out.close() ;
37
38    cout << "Copy completed. "
39         << count - 1 << " characters copied." << endl ;
40    return 0 ;
41  }
```

After opening both files, the program reads the first character of the input file on line 26. The loop on lines 28 to 33 continues to write a character to the output file and read the next character from the input file until the end of the input file is detected.

A message displaying the number of characters copied is displayed on lines 38 and 39.

The program can be improved by using an easier while statement than that used on line 28:

```
while ( in.get( c ) )
```

Evaluates to true *if data read from the file; otherwise evaluates to* false

The improvement is made in the next program. Note that there is no special processing necessary for the first character as required in program P14C (line 26).

Program Example P14D

```
1   // Program Example P14D
2   // Program to demonstrate get, put and end of file detection.
3   #include <iostream>
4   #include <fstream>
5   using namespace std ;
6
7   int main( void )
8   {
9     char c ;
10    int count = 0 ;
11
12    ifstream in( "file.txt" ) ;   // Open the input file.
13    if ( in.fail() )
14    {
15      cerr << "Open failure on file.txt" << endl ;
16      return 1 ;
17    }
18
19    ofstream out( "file.bak" ) ; // Open the output file.
20    if ( out.fail() )
21    {
22      cerr << "Open failure on file.bak" << endl ;
23      return 1 ;
24    }
25
26    while( in.get( c ) ) // Read a character while not end of file.
27    {
28      out.put( c ) ; // Write the character.
29      count++ ;
30    }
31
32    in.close() ;
33    out.close() ;
34    cout << "Copy completed. "
35         << count << " characters copied." << endl ;
36    return 0 ;
37  }
```

The same technique for detecting the end of file can be applied to any member function that returns a reference to a stream. The next program demonstrates this by reading a file "word" by "word" using the stream extraction operator >>.

这种检测文件末尾的方法同样可以应用于返回流引用的成员函数。下面的程序使用流提取运算符>>逐个单词地读取文件words.txt，演示了这个方法。

Program Example P14E

```
1   // Program Example P14E
2   // Program to demonstrate end of file detection when using >>.
3   #include <iostream>
4   #include <fstream>
5   #include <iomanip>
6   #include <string>
7   using namespace std ;
8
9   int main( void )
10  {
11    string word ;
12    int word_count = 0 ;
13
14    ifstream in( "words.txt" ) ;
15    if ( in.fail() )
16    {
17      cerr << "Open failure on words.txt" << endl ;
18      return 1 ;
19  }
20
21    while ( in >> word ) // Read a word while not end of file.
22    {
23      cout << word << endl ;
24      word_count++ ;
25    }
26
27    in.close() ;
28    cout << "Number of words read = " << word_count << endl ;
29    return 0 ;
30  }
```

The program assumes that each "word" is delimited by one or more whitespace characters. Whitespace characters are ignored by >>.

The file is opened on line 14. If the file cannot be opened an error message is displayed and 1 is returned to the operating system indicating an unsuccessful completion of the program.

The while loop from line 21 to line 25 continues to display each "word" in the file until the end of the file is detected. When the end of the file occurs, the while loop terminates, the file is closed and the number of "words" in the file is displayed. The program finishes by returning 0 to the operating system to indicate a successful completion of the program.

这个程序假设每个单词之间用一个或者多个空白字符来分隔。空白字符会被>>忽略掉。

第14行打开文件，如果文件打开失败，则显示一条错误信息，并向操作系统返回1，表示程序异常终止。

第21行至第25行的while循环将文件内容连续显示在屏幕上，直到文件尾为止。到达文件尾以后，while循环结束，将文件关闭，并显示文件中的单词总数。然后程序向操作系统返回0，表示程序正常结束。

14.5　Appending data to the end of a file
（向文件末尾添加数据）

To append data to the end of a file, the file mode must be specified when opening the file. The file mode values are specified in the `ios` class. The possible file mode values are shown in Table 14.3.

为了向一个文件的末尾添加数据，在打开文件的时候必须指定文件的打开模式。文件的打开模式在ios类中定义，如表14.3所示。

Table 14.3　File modes

Mode	Meaning
`ios::app`	Opens a file for appending - additional data is written at the end of the file.
`ios::ate`	Start at the end of the file when a file is opened.
`ios::binary`	Opens a file in binary mode (default is text mode).
`ios::in`	Opens a file for input (default for `ifstream` objects).
`ios::out`	Opens a file for output (default for `ofstream` objects).
`ios::trunc`	Truncates (deletes) the file contents if the file already exists.
`ios::nocreate`	Do not create a new file. Open fails if the file does not already exist. For output files only.
`ios::noreplace`	Do not replace an existing file. Open fails if the file already exists. For output files only.

These values may be combined using the bitwise operator OR (|). For example, to open a file for input and output:

```
fstream in_out( "file.txt", ios::in | ios::out ) ;
```

The open mode of a file was not specified in the previous programs. This is because the default open mode for `ifstream` objects is `ios::in` and `ios::out` for `ofstream` objects.

The next program demonstrates appending the line "This line is added to the end of the file" to the file `file.txt`.

可以通过使用按位或运算符 | 来联合使用多种打开模式。例如，下面的语句表明打开一个文件，使其既可用作输入，也可用作输出。在前面的程序中，都没有指定文件的打开模式。这是因为ifstream类对象的默认打开模式是ios::in，而ofstream 类对象的默认打开模式是ios::out。

Program Example P14F

```
1   // Program Example P14F
2   // Program to demonstrate adding data to the end of a file.
3   #include <iostream>
4   #include <fstream>
5   using namespace std ;
6
7   int main( void )
8   {
9     ofstream out( "file.txt", ios::app ) ;
10    if ( out.fail() )
11    {
12      cerr << "Open failure on file.txt" << endl ;
13      return 1 ;
14    }
15    out << "This line is added to the end of the file" << endl ;
```

```
16    out.close() ;
17    return 0 ;
18  }
```

14.6 Reading lines from a file（从文件中读取行）

An entire line can be read from a file into either a C-string or a C++ string using the function `getline()`. Program P14G uses a C++ string to store a line, while P14H uses a C string for the same purpose. Both programs read each line of the source file and display the line on the screen preceded by a line number.

使用函数getline( )可以从文件读入一整行数据到C/C++字符串中。程序P14G使用C++字符串存储从文件中读入的一行数据，而程序P14H则使用C字符串存储。这两个程序都是逐行读取源文件，并将其显示在屏幕上，显示时增加了前导的行号。

Program Example P14G

```
1   // Program Example P14G
2   // Program to demonstrate reading a file line by line
3   // using a C++ style string.
4   #include <iostream>
5   #include <fstream>
6   using namespace std ;
7
8   int main( void )
9   {
10    string cpp_string ; // A C++ style string.
11    int line_number = 0 ;
12
13    ifstream in( "p14g.cpp" ) ; // Open this program file.
14    if ( in.fail() )
15    {
16      cerr << "Open failure on p14g.cpp" << endl ;
17      return 1 ;
18    }
19    // Read each line into a C++ string,
20    // and display the line number and the string.
21    while ( getline( in, cpp_string) )
22    {
23      cout << ++line_number << ":" << cpp_string << endl ;
24    }
25    in.close() ;
26    return 0 ;
27  }
```

Program Example P14H

```
1   // Program Example P14H
2   // Program to demonstrate reading a file line by line
3   // using a C-style string.
4   #include <iostream>
5   #include <fstream>
6   using namespace std ;
7
8   int main( void )
```

```
9   {
10    char c_string[81] ; // A C-style string, maximum size is 80.
11    int line_number = 0 ;
12
13    ifstream in( "p14h.cpp" ) ;
14    if ( in.fail() )
15    {
16      cerr << "Open failure on p14h.cpp" << endl ;
17      return 1 ;
18    }
19    // Read each line into a C-string
20    // and display the line number and the string.
21    while( in.getline ( c_string, 81 ) )
22    {
23      cout << ++line_number << ':' << c_string << endl ;
24    }
25    in.close() ;
26    return 0 ;
27  }
```

14.7 Random access（随机存取）

The previous programs all perform *serial file processing*. With serial file processing, data items are read or written one after the other. For example, if you wanted to read the fifth data item in a file, with serial processing you must read the previous four data items first. With *random* or *direct access* you can move around in a file, reading and writing at any position in the file.

The position in a file is held in the file position marker (FPM).

The istream member function seekg() (meaning 'seek get') is used to set the FPM for an input file, from where the next input data is read. This function requires two arguments. The first argument is an offset that tells the FPM how many bytes to skip. The second argument specifies the point from where the offset is measured and has three possible values as shown in Table 14.4.

前面的所有程序执行的都是顺序文件处理。在顺序文件处理过程中，对数据项的读/写是一个接着一个进行的。例如，如果想读取文件中的第五个数据项，那么使用顺序处理方法必须先读取前四个数据项才能读取第五个数据项。但是随机存取（或称直接存取）允许在文件中随意定位，并在文件的任何位置读写数据。

文件中的位置信息保存在文件位置标记（FPM）中。

istream 类的成员函数 seekg( ) 用来为输入文件设定 FPM，指示下一个要读取的数据的位置。这个函数需要两个实参，第一个实参是一个偏移量，它告诉 FPM 要跳过多少字节，第二个实参确定偏移量计算的起始位置，其可能取值有 3 种，如表 14.4 所示。

Table 14.4   Offset values

ios::beg (=0)	The offset is from the beginning of the stream.
ios::cur (=1)	The offset is from the current position in the stream.
ios::end (=2)	The offset is from the end of the stream.

Examples:

Open a file containing the characters A to J with

```
istream in( "letters.dat" ) ;
```

When the file is opened the FPM is set to 0.

```
         | A | B | C | D | E | F | G | H | I | J |
FPM = 0 ⌐
```

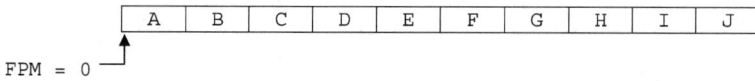

```
  in.seekg( 4, ios::beg ) ;  // Move the FPM past the 4th. character.
or
  in.seekg( 4 ) ;            // ios::beg can be omitted.
```

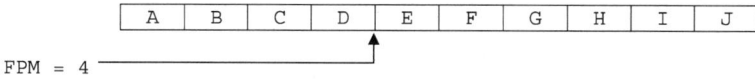

```
         | A | B | C | D | E | F | G | H | I | J |
FPM = 4 _____↑
```

```
  in.seekg( 1, ios::cur ) ;  // Move the FPM forward 1 character.
```

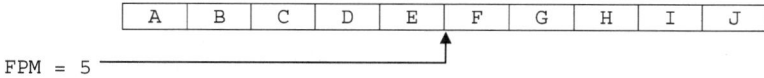

```
         | A | B | C | D | E | F | G | H | I | J |
FPM = 5 _____↑
```

```
  in.seekg( -3, ios::cur ) ; // Move the FPM back 3 characters.
```

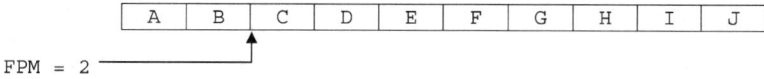

```
         | A | B | C | D | E | F | G | H | I | J |
FPM = 2 _____↑
```

```
  in.seekg( 0, ios::beg ) ;  // Move the FPM to the start of the file.
```

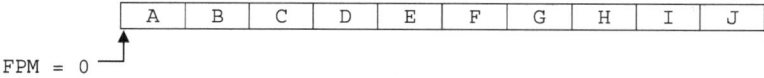

```
         | A | B | C | D | E | F | G | H | I | J |
FPM = 0 ⌐
```

```
  in.seekg( 0, ios::end ) ;  // Move the FPM to the end of the file.
```

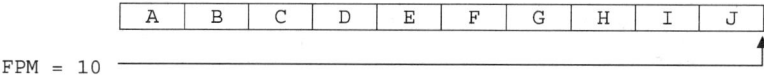

```
         | A | B | C | D | E | F | G | H | I | J |
FPM = 10 _____↑
```

The equivalent function for setting the FPM for an output file is the seek put function, seekp().

用于在输出文件中设置FPM的函数是seekp( )。

There are also two other functions tellg() and tellp() that return the current value of the FPM for input and output streams respectively.

还有两个函数tellg( )和tellp( )分别用来返回输入流和输出流中FPM的当前值。

The next program is a simple demonstration of seekg() and tellg(). In this program, after the user enters a file position, the character at that position and its ASCII value are displayed.

Program Example P14I

```
1  // Program Example P14I
2  // Simple demonstration of random file access.
3  #include <iostream>
4  #include <fstream>
5  using namespace std ;
6
7  int main( void )
```

```
8  {
9     char c ;
10    int file_pos, last_pos, offset ;
11
12    ifstream in( "letters.txt" ) ;
13    if ( in.fail() )
14    {
15      cerr << "Open failure on letters.txt" << endl ;
16      return 1 ;
17    }
18
19    // Find the position of the last character in the file.
20    in.seekg( 0, ios::end ) ;
21    last_pos = in.tellg() ;
22
23    // Loop until user enters a 0.
24    do
25    {
26      cout << "Enter the file position (0 to end) " ;
27      cin >> file_pos ;
28      // Is this a valid position?
29      if ( file_pos > last_pos || file_pos < 0 )
30        cout << "Invalid position. Enter a value between 1 and "
31             << last_pos << endl ;
32      if ( file_pos > 0 && file_pos <= last_pos )
33      {
34        // The offset is 1 less than the position.
35        offset = file_pos - 1 ;
36        // Go directly to the character's offset.
37        in.seekg( offset, ios::beg ) ;
38        // Read the character and display.
39        in.get( c ) ;
40        cout << "Character at position " << file_pos
41             << ", offset " << offset
42             << " is "<< c << " ASCII "
43             << static_cast<int>( c ) << endl ;
44      }
45    }
46    while ( file_pos != 0 ) ;
47    in.close() ;
48    return 0 ;
49  }
```

As sample run of this program follows:

```
Enter the file position (0 to end) 3
Character at position 3, offset 2 is C ASCII 67
Enter the file position (0 to end) 0
```

The seekg() function on line 20 moves the FPM to the end of the file. The FPM is now positioned just after the last character in the

第20行的函数seekg( )将FPM移到文件的末尾。现在FPM指向文件中最后一个字符后面的位置。第21行的函数tellg( )读取FPM的值，此时FPM的值也代表文件中字符的个数。

第27行请用户输入想要读取的文

file. The `tellg()` function on line 21 reads the value of the FPM, which is also the number of characters in the file.

Line 27 inputs the required position in the file, which is converted to an offset on line 35.

The `seekg()` function on line 37 moves the FPM to just before the character to be read. The character is read on line 39 and displayed along with its ASCII value on lines 40 to 43.

Note the use of casting on line 43 to convert the value of `c` to its equivalent integer value.

件位置，这个值在第35行被转换成偏移量。

第37行 的 函数seekg( )将FPM移到将要读取的字符的前面。第39行读取该字符，然后在第40行至第43行将该字符及其ASCII码值一起显示到屏幕上。

注意第43行使用强制类型转换将变量c的值转换成与其等价的整数，即ASCII码值。

14.8   Object I/O（对象 I/O）

In program P10G the `>>` and `<<` operators were overloaded for keyboard input and screen output operations specific to the `time24` class. No modifications to the `time24` class are needed to use disk files instead of the keyboard and screen.

The next program opens a file and writes and reads `time24` objects to and from this file.

在程序P10G中，通过重载运算符>>和<<使类time24具有键盘输入和屏幕输出的功能。使用磁盘文件来代替键盘和屏幕进行输入和输出，并不需要对time24类进行修改。

Program Example P14J

```
1   // Program Example P14J
2   // Demonstration of object I/O.
3   #include <iostream>
4   #include <fstream>
5   #include "time24.h"
6   #include "time24.cpp"
7   using namespace std ;
8
9   int main( void )
10  {
11    time24 t1( 1, 2, 3 ) ;
12    time24 t2( 10, 10, 10 ) ;        Both input and output.
13
14    fstream in_out( "times.dat", ios::in|ios::out ) ;
15    if ( in_out.fail() )
16    {
17      cerr << "error on opening times.dat" << endl ;
18      return 1 ;
19    }
20
21    // Write objects to the file.
22    in_out << t1 << t2 ;
23
24    // Rewind the file to the start.
25    in_out.seekg( 0 ) ;
26
27    // Read objects back and display on the screen.
28    in_out >> t1 >> t2 ;
```

```
29    cout << t1 << t2 ;
30    in_out.close() ;
31    return 0 ;
32 }
```

The output from this program is:

```
01:02:03
10:10:10
```

The file `time24.h` included on line 5 contains the `time24` class declaration from program P10H.

Line 6 includes the file `time24.cpp` containing the `time24` class member functions, also from program P10H.

Line 14 opens the file `times.dat` as an `fstream` object for input and output.

Line 22 writes the `time24` objects `t1` and `t2` to the file.

Line 28 reads the objects back from the file and line 29 displays the objects.

14.9   Binary I/O（二进制 I/O）

There are two types of files in C++: *text* (or *ASCII*) files and *binary* files. The difference between the two types of files is in the way they store numeric data types. In binary files, numeric data is stored in binary format, while in text files numeric data is stored as ASCII characters. For example, consider the following definition of an integer variable n:

C++文件有两种类型：文本文件（或者ASCII文件）和二进制文件。它们的差别在于存储数值型数据的方式不同。在二进制文件中，数值型数据是以二进制形式存储的，而在文本文件中，则是以字符的ASCII码形式存储的。

```
short int n = 123 ;
```

From program P2D, the variable n occupies two bytes of memory. However, storing the value of n in a text file requires three bytes of memory:

由程序P2D可知，短整型变量n在内存中占两个字节，但把变量n的值存储在文本文件中则需要3个字节的内存。

Character :	'1'	'2'	'3'
ASCII value in decimal:	49	50	51
ASCII value in binary:	00110001	00110010	00110011

Each digit of the number requires one byte of storage in an ASCII file. In a binary file, the digits of a number do not occupy individual storage locations. Instead the number is stored in its entirety as a binary number. So the variable n, with a value of 123, will be stored in two bytes as:

在ASCII文件中，每一位数字都要占用一个字节的存储空间。在二进制文件中，一个数据的每一位数字并不占用单独的存储单元，而是把整个数据作为一个二进制数来存储。

00000000	01111011

If, for example, the value of n is increased to 1234, the binary file will still require the same storage as it does for the number 123. However, an ASCII file would require one more byte to store the extra digit.

举例来说，如果把n值增大到1234，那么对于二进制文件而言，存储1234和存储123需要相同大小的存储空间；而对于ASCII码文件来说，则需要再增加一个字节来存储额外的数字4。

14.9.1   Serial writing of objects to a binary file

The two-argument `istream` member function `write()` is used to write an object in binary format. The first argument is the address of the block of memory where the object is stored and the second argument is the size of the block in bytes.

It is important to note that the object must be stored in one contiguous block of memory. This means that the object cannot have a pointer data member to dynamically allocated memory. It also means that the class cannot have any `string` data members, since the `string` class uses a pointer data member.

The next program demonstrates `write()` by writing stock objects to a file. The stock class uses a fixed length C-string to store the stock description.

带有两个实参的istream类的成员函数write( )用于以二进制格式编写一个对象。第一个实参是存储对象的内存区域的地址，第二个实参是这个内存区域的字节大小。注意：对象必须存储在一块连续的内存区域中，这就意味着对象不能有指向动态分配的内存的指针数据成员；这也意味着类不能有任何string数据成员，因为string类使用一个指针数据成员。

Program Example P14K

```
1   // Program Example P14K
2   // Demonstration of writing to a binary file.
3   #include <iostream>
4   #include "stock.h"
5   #include "stock.cpp"
6   using namespace std ;
7
8   int main( void )
9   {
10    // Initialise an array of stock objects.
11    class stock stationery[ 10 ] =
12    { stock( 1, "Pencil", 68 ),
13      stock( 2, "Pen", 30 ),
14      stock( 3, "Highlighter", 90 ),
15      stock( 4, "Eraser", 24 ),
16      stock( 5, "Pencil Sharpener", 5 ),
17      stock( 6, "Pocket Folder", 50 ),
18      stock( 7, "Paper Tie", 300 ),
19      stock( 8, "Glue Stick", 30 ),
20      stock( 9, "Box File", 35 ),
21      stock( 10, "Note Book", 97 ) } ;
22
23    int object_size = sizeof( stock ) ;
24
25    // Open the stock file for binary output.
26    ofstream stock_file( "stock.dat", ios::binary ) ;
27    if ( stock_file.fail() )
28    {
29      cerr << "Open failure on stock.dat" << endl ;
30      return 1 ;
31    }
32
```

```
33   // Write each object to the output file.
34   for ( int i = 0 ; i < 10 ; i++ )
35   {
36     cout << "Writing Stock Code "
37          << stationery[ i ].get_code() << ' '
38          << stationery[ i ].get_description()
39          << " quantity = " << stationery[ i ].get_quantity()
40          << endl ;
41     stock_file.write( reinterpret_cast<char *>( &stationery[i] ),
42                       object_size ) ;
43   }
44   stock_file.close() ;
45   return 0 ;
46 }
```

The output from this program is:

```
Writing Stock Code 1 Pencil quantity = 68
Writing Stock Code 2 Pen quantity = 30
Writing Stock Code 3 Highlighter quantity = 90
Writing Stock Code 4 Eraser quantity = 24
Writing Stock Code 5 Pencil Sharpener quantity = 5
Writing Stock Code 6 Pocket Folder quantity = 50
Writing Stock Code 7 Paper Tie quantity = 300
Writing Stock Code 8 Glue Stick quantity = 30
Writing Stock Code 9 Box File quantity = 35
Writing Stock Code 10 Note Book quantity = 97
```

The file stock.h contains:

```
1  #if !defined STOCK_H
2  #define STOCK_H
3
4  // Header for a simple stock class.
5  #include <fstream>
6  using namespace std ;
7
8  class stock
9  {
10 public:
11   stock() ;
12   // Purpose: Default constructor.
13
14   stock( int code, char desc[ ], int quantity ) ;
15   // Purpose   : Constructor
16   // Parameters: stock code, description and stock quantity.
17   //           : Note that the description is a C-string.
18
19   void set_quantity( int quantity ) ;
20   // Purpose  : Assign a value to the stock quantity.
21   // Parameter: The quantity to be assigned.
22
```

```
23    const int get_code ( void ) const ;
24    // Purpose: Inspector function to return the stock code.
25
26    const char* get_description( void ) const ;
27    // Purpose: Inspector function to return the stock description.
28    //         : Note that description is a C-string.
29
30    const int get_quantity ( void ) const ;
31    // Purpose: Inspector function to return the stock quantity.
32
33    void binary_write( ofstream& os );
34    // Purpose  : Write stock object to a binary file.
35    // Parameter: os - File output stream.
36
37    bool binary_read( ifstream& is ) ;
38    // Purpose  : Read stock object from a binary file.
39    // Parameter: is - File input stream.
40    // Returns  : true if end of file, otherwise false is returned.
41
42 private:
43    int stock_code ;
44    char description[21] ;
45    int quantity_in_stock ;
46 } ;
47
48 #endif
```

The file stock.cpp contains:

```
1   #if !defined STOCK_CPP
2   #define STOCK_CPP
3
4   // Member function definitions for stock class.
5   #include "stock.h"
6   #include <cstring>
7   stock::stock()
8   {
9     stock_code = 0 ;
10    quantity_in_stock = 0 ;
11    description[0] = '\0' ;
12  }
13
14  stock::stock( int code, char desc[ ], int quantity )
15  {
16    stock_code = code ;
17    strcpy( description, desc ) ;
18    quantity_in_stock = quantity ;
19  }
20
21  void stock::set_quantity ( int quantity )
22  {
```

```
23   quantity_in_stock = quantity ;
24 }
25
26 const int stock::get_code ( void ) const
27 {
28   return stock_code ;
29 }
30
31 const char* stock::get_description( void ) const
32 {
33   return description ;
34 }
35
36 const int stock::get_quantity( void ) const
37 {
38   return quantity_in_stock ;
39 }
40
41 void stock::binary_write( fstream& os )
42 {
43   os.write( reinterpret_cast<char *>( this ), sizeof( *this ) ) ;
44 }
45
46 bool stock::binary_read( fstream& is )
47 {
48   is.read( reinterpret_cast<char *>( this ), sizeof( *this ) );
49   return !is.eof() ;
50 }
51
52 #endif
```

The write() function on line 41 of program P14K treats the stock object as a block of memory bytes or characters, which it copies from the memory to a disk file without converting to ASCII.

Since the object is treated as a block of characters, the first argument in write() is a pointer to a character string. This is why the address of the ith stock object, &stationery[i], is cast to a pointer to a character string by

```
reinterpret_cast<char*>( &stationery[i] )
```

Unlike the static_cast, introduced in chapter two, the reinterpret_cast allows any pointer type to be converted to any other pointer type.

The second argument in write() is the number of bytes in the memory block, which is calculated on line 23 of program P14K.

第41行的函数write( )把stock对象看作一整块的内存字节或字符，无须转换为ASCII码就可直接从内存将其复制到磁盘文件中。

既然对象被看作字符块，那么函数write( )的第一个实参就应该是一个指向字符串的指针。这就是第i个stock对象的地址&stationery[i]被强制转换成指向字符串的指针的原因。

与第2章介绍的static_cast不同的是，reinterpret_cast允许从任意指针类型转换到其他任意指针类型。函数write( )的第二个实参是内存块的字节数，它由程序P14K的第23行来计算。

14.9.2   Serial reading of objects from a binary file

The `istream` member function `read()` is used to read an object's data from a binary file into memory. This function uses the same arguments as `write()`, as demonstrated in the next program.

istream类的成员函数read( )用于从二进制文件中将对象的数据读到内存中。函数read( )的实参和函数write( )的相同，下面的程序演示了这个函数的用法。

Program Example P14L

```
1  // Program Example P14L
2  // Demonstration of serial reading of a binary file.
3  #include <iostream>
4  #include <iomanip>
5  #include "stock.h"
6  #include "stock.cpp"
7  using namespace std ;
8
9  int main( void )
10 {
11 stock stock_record ;
12
13   // Open the stock file for binary input.
14   ifstream stock_file( "stock.dat", ios::binary ) ;
15   if ( stock_file.fail() )
16   {
17     cerr << "Open failure on stock.dat" << endl ;
18     return 1 ;
19   }
20
21   int object_size = sizeof( stock ) ;
22
23   cout << "Code Description Quantity" << endl ;
24   cout << left ; // Left justify data.
25   // Read and display stock records until eof.
26   while( stock_file.read
27          ( reinterpret_cast<char *>( &stock_record ),
28            object_size ) )
29   {
30     cout << setw(5) << stock_record.get_code()
31          << setw(20) << stock_record.get_description()
32          << stock_record.get_quantity() << endl ;
33   }
34   stock_file.close() ;
35   return 0 ;
36 }
```

The output from this program is:

```
Code Description        Quantity
1    Pencil             68
2    Pen                30
3    Highlighter        90
4    Eraser             24
```

5	Pencil Sharpener	5
6	Pocket Folder	50
7	Paper Tie	300
8	Glue Stick	30
9	Box File	35
10	Note Book	97

14.9.3   Binary I/O as class member functions

Binary I/O is made easier for the users of a class (the client programmers) if the details of the read and write functions are "hidden" in class member functions.

如果将读/写函数的细节隐藏在类的成员函数中，那么程序员使用二进制I/O就比较容易了。

The file `stock.cpp` contains the following member function to write to a binary file:

```
void stock::binary_write( fstream& os ) const
{
  os.write( reinterpret_cast<char *>( this ), sizeof( *this ) ) ;
}
```

Using this member function in program P14K, lines 41 and 42 become:

```
stationery[i].binary_write( stock_file ) ;
```

The output stream `stock_file` is passed to the parameter `os` of the member function `binary_write()`. The built-in pointer `this` is a pointer to the object that invoked the member function, i.e. `stationery[i]`. The pointer `this` is therefore equivalent to `&stationery[i]` and `sizeof( *this )` is equivalent to `sizeof ( stationery[i] )`. The file `stock.cpp` also contains the following member function to read from a binary file:

```
bool stock::binary_read( fstream& is )
{
  is.read( reinterpret_cast<char *>( this ), sizeof( *this ) ) ;
  return !is.eof() ;
}
```

Using this member function in program P14L, lines 26 to 28 become:

```
while ( stock_record.binary_read( stock_file ) )
```

In addition to reading an object from the input stream, `binary_read()` returns `false` if the end of the file has been reached, otherwise it returns `true`. The while loop will, therefore, continue until the end of the file is reached.

除了从输入流中读取对象，函数binary_read( )将在到达文件末尾时返回false，否则返回true。因此，while循环将一直执行到文件尾时为止。

There is one more modification required in program P14L, if the program is to successfully compile. Line 14 must be changed to:

```
fstream stock_file( "stock.dat", ios::binary | ios::in ) ;
```

The reason for this is to make stock_file an `fstream` object as required by the function bool `stock::binary_read()`.

14.9.4   Binary file random access

The next program demonstrates the direct or random access of a stock object stored in a binary file. The layout of the file as created by program P14K is illustrated in the diagram below.

下面的程序演示了如何直接访问存储在二进制文件中的stock对象。由程序P14K创建的文件结构如下图所示。

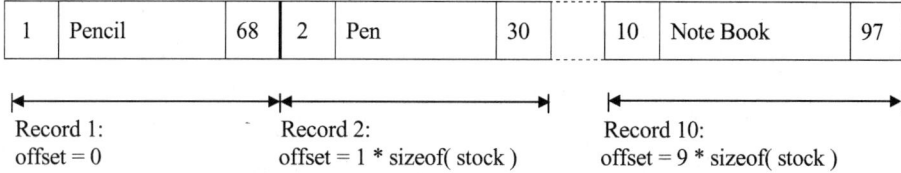

This type of file is called a *relative* file. In this file structure, the stock code is related to the position of the record in the file, i.e. stock code 1 is the first record, stock code 2 is the second record and stock code 3 is the third record and so on. The offset of a stock record in the file can be calculated by subtracting 1 from the stock code and multiplying the result by the size of the stock record.

The following program allows the user to enter a stock code and updates the corresponding stock quantity. The user can update the stock records in any order. After the updates are complete, the program displays the contents of the stock file.

这种类型的文件称为相对文件。在这种文件结构中，存货代码和存货记录在文件中的位置有关，即存货代码1为第1条记录，存货代码2为第2条记录，存货代码3为第3条记录，以此类推。

一个存货记录在文件中的偏移量可以用存货代码值减去1，然后乘以存货记录的字节大小来计算。

Program Example P14M

```
1   // Program Example P14M
2   // Program to demonstrate random access reading of a binary file.
3   #include <iostream>
4   #include <iomanip>
5   #include "stock.h"
6   #include "stock.cpp"
7   using namespace std ;
8
9   // Non-class standalone functions.
10  int input_stock_code( void ) ;
11  // Purpose: Function to input a stock code from the keyboard.
12  // Returns: The stock code entered.
13
14  void update_stock_record( int stock_code, fstream& stock_file ) ;
15  // Purpose   : Update the stock quantity.
16  // Parameters: The stock code and the output stream.
17
18  int main( void )
19  {
20    int stock_code, max_stock_code ;
21    stock stock_record ;
22
```

```
23   // Open the stock file for binary input and output.
24   fstream stock_file( "stock.dat",
25                       ios::binary | ios::in | ios::out) ;
26   if ( stock_file.fail() )
27   {
28     cerr << "Open failure on stock.dat" << endl ;
29     return 1 ;
30   }
31
32   // Find the number of records in the file.
33   stock_file.seekg( 0, ios::end );
34   max_stock_code = stock_file.tellg() / sizeof( stock ) ;
35
36   // Input a stock code and update the stock quantity, code 0 ends.
37   do
38   {
39     stock_code = input_stock_code() ;
40     if ( stock_code > 0 && stock_code <= max_stock_code )
41       update_stock_record( stock_code, stock_file ) ;
42   }
43   while ( stock_code != 0 ) ;
44
45   // Display the updated file.
46   cout << "Code Description        Quantity" << endl ;
47   cout << left ; // Left justify data.
48   // Read and display stock records until end of the file.
49   stock_file.seekg( 0 ) ;
50   while ( stock_record.binary_read( stock_file) )
51   {
52     cout << setw(5) << stock_record.get_code()
53          << setw(20)<< stock_record.get_description()
54          << stock_record.get_quantity() << endl ;
55   }
56
57   stock_file.close() ;
58   return 0 ;
59 }
60
61 int input_stock_code( void )
62 {
63   int code ;
64   cout << "Enter a stock code (0 to end ): " ;
65   cin >> code ;
66   return code ;
67 }
68
69 void update_stock_record( int stock_code, fstream& stock_file )
70 {
71   int quantity ;
72   int offset ;
```

```
73   stock stock_record ;
74
75   // Place the file position marker just before the record.
76   offset = ( stock_code -1 ) * sizeof( stock_record ) ;
77   stock_file.seekg( offset, ios::beg ) ;
78
79   // Read the record.
80   stock_record.binary_read( stock_file ) ;
81
82   // Get the new quantity.
83   cout << "Enter Stock Quantity for Stock Code: "
84       << stock_code << ", "
85       << stock_record.get_description() << ": ";
86   cin >> quantity ;
87
88   // Update the record.
89   stock_record.set_quantity( quantity ) ;
90   stock_file.seekp( offset, ios::beg ) ;
91   stock_record.binary_write( stock_file ) ;
92 }
```

A sample run of this program is:

```
Enter a stock code (0 to end ): 3
Enter Stock Quantity for Stock Code: 3, Highlighter: 12
Enter a stock code (0 to end ): 6
Enter Stock Quantity for Stock Code: 6, Pocket Folder: 22
Enter a stock code (0 to end ): 1
Enter Stock Quantity for Stock Code: 1, Pencil: 50
Enter a stock code (0 to end ): 9
Enter Stock Quantity for Stock Code: 9, Box File: 35
Enter a stock code (0 to end ): 0
Code Description          Quantity
1    Pencil               50
2    Pen                  30
3    Highlighter          12
4    Eraser               24
5    Pencil Sharpener     5
6    Pocket Folder        22
7    Paper Tie            300
8    Glue Stick           30
9    Box File             35
10   Note Book            97
```

The program continually asks the user for a stock code and quantity until the user enters a stock code of 0.

The stock record corresponding to the entered stock code is retrieved by calculating the offset of the record in the file, moving the file position marker to the start of the record, and reading the record. This is done on lines 76 to 80.

程序将不断请求用户输入存货代码和数量，直到用户输入的存货代码等于0时为止。

与输入的存货代码相对应的存货记录可以通过计算该记录在文件中的偏移量来获取，即将文件位

After updating the stock quantity in memory with the member function `set_quantity()`, it is necessary to write the updated record back to the file. This is done by first moving back to the start of the record (line 90) and then writing the updated record (line 91).

After the updating of the file is complete, the entire file is displayed. This is done by first moving back to the start of the file on line 49, and then displaying each stock record on lines 50 to 55.

置标记（FPM）移到该记录的起始位置，然后读取该记录。这些工作是由程序的第76行至第80行完成的。

在使用成员函数set_quantity( )对内存中的存货数量进行更新以后，还要将更新的记录写回到文件中，即先将FPM移到该记录的起始位置（见第90行），然后将更新的记录写入文件（见第91行）。

当文件中的数据更新完毕之后，显示整个文件的内容。首先在第49行将FPM 移到文件的起始位置，然后通过第50行至第55行的程序显示每一个存货记录。

Programming pitfalls

1. File names frequently contain \, for example, `c:\newfile.dat`. To open such a file you have to use an extra \ as in:

```
istream in( "c:\\newfile.dat" ) ;
```

Without the extra \, the `\n` is interpreted as the newline character. Linux and UNIX uses / in file specifications, and so the problem does not arise.

1. 文件名通常都会包含反斜杠\，例如c:\newfile.dat。为了打开这样一个文件，必须使用一个额外的\，就像下面这样。
如果没有这个额外的\，那么\n将会被解释为转义序列中的换行符。UNIX和Linux在文件说明中使用正斜杠/，所以不会出现这个问题。

2. In the `istream` member function `read()`, the first argument is a character pointer to a memory location. The memory location must be big enough to store the data to be read.

2. 在istream类的成员函数read( )中，第一个实参是一个指向内存单元的字符指针。该内存单元必须足够大，以便能够存储将要读入的数据。

3. The default open mode of an `ifstream` object is `ios::in` and the default mode for an `ofstream` object is `ios::out`. No default mode exists for an `fstream` object and must be specified, e.g.

```
fstream in_out( "file.txt", ios:in | ios::out ) ;
```

3. ifstream类对象的默认打开模式为ios::in，而ofstream类对象的默认打开模式为ios::out。fstream类对象没有定义默认打开模式，所以在打开一个fstream对象时，必须指定文件的打开模式。

4. The open mode values may be combined using the bitwise OR operator |, not the logical OR operator ||.

4. 使用按位或运算符 | 可以将多种文件打开模式联合使用，但不能使用逻辑或运算符 ||。

Quick syntax reference

	Syntax	Examples

Open a file for input	`ifstream stream_name( filename, mode ) ;`	`ifstream in( "file" ) ;`
		`if ( in.fail() )`
		`{`
		`  cerr << "file error" ;`
		`  return 1 ;`
		`}`

Open a file for output	`ofstream stream_name( filename, mode ) ;`	`ofstream out( "file" ) ;`
		`if ( out.fail() )`
		`{`
		`  cerr << "file error" ;`
		`  return 1 ;`
		`}`

Open a file for input and output	`fstream stream_name( filename, mode ) ;`	`fstream in_out( "file",`
		`            ios::in|ios::out ) ;`
		`if ( in_out.fail() )`
		`{`
		`  cerr << "file error" ;`
		`  return 1 ;`
		`}`

(cont.)

	Syntax	Examples
Read a character from a file	`stream_name.get( variable ) ;`	`in.get( ch ) ;`
Write a character to a file	`stream_name.put( variable ) ;`	`out.put( ch ) ;`

Read a C++ string from a file	`getline( stream_name,`
	 `cpp_string_variable ) ;`
	 `or`
	 `>>`
		`string cpp_str ;`
		 `getline ( in, cpp_str ) ;`
		 `in >> cpp_str ;`

Read a C-string from a file	`stream_name.get (`
	 `    c_string_variable,`
	 `    max_number_of_characters,`
	 `    delimiter ) ;`
	 `or`
	 `stream_name.read(`
	 `    c_string_variable,`
	 `    number_of_characters ) ;`
	 `or`
	 `>>`
		`char c_str[81] ;`
		
		 `in.get( c_str, 81, '\n' ) ;`
		
		 `in.read( c_str, 81 ) ;`
		
		 `in >> c_str ;`

Write a C-string to a file	`steam_name.write(`
	 `    c_string, variable`
	 `    number_of_characters ) ;`
		`char c_str[]="ABC";`
		 `out.write( c_str, 3 ) ;`

Set the file position marker for an input file	`stream_name.seekg( offset,`
	 `                    origin ) ;`
	 `// offset = integer value`
	 `// origin = ios:beg,`
	 `//          ios:cur or`
	 `//          ios:end`
		`int off = 0 ;`
		 `in.seekg( off, ios:beg ) ;`

Set the file position marker for an output file	`stream_name.seekp( offset,` `                    origin ) ;`	`int off = 0 ;` `in.seekp( off, ios:end ) ;`
Closing a file	`stream_name.close() ;`	`out.close() ;`

Binary file block input	`stream_name.read(`
	 `        character_pointer,`
	 `        number_of_characters ) ;`
		`in.read(`
		 `reinterpret_cast<char *>`
		 `( &stock_record),`
		 `    sizeof( stock_record ) ) ;`

Binary file block output	`stream_name.write(`
	 `        character_pointer,`
	 `        number_of_characters ) ;`
		`out.write(`
		 `reinterpret_cast<char *>`
		 `( &stock_record ),`
		 `    sizeof( stock_rec ) ) ;`

Exercises

1. Write C++ statements to open the following files:

File name	Mode	File type
supplier.dat	input	binary
customer.dat	input and output	binary
temp.txt	output	text
file.txt	append	text

2. Which mode would you use for the following?

    (a) updating existing data in a file

    (b) appending new data to the end of a file

    (c) deleting the contents of an existing file before writing new data to it.

3. Write a program to count the number of characters and the number of words in a text file.

4. Write a program to remove blank and zero length lines from a text file. An input file and an output file are required.

5. Write a program to compare two text files and display any differences between the files.

6. What does the file `alpha.txt` contain after this program is run?

```cpp
#include <iostream>
#include <fstream>
using namespace std ;

int main( void )
{
  char c ;
  fstream alpha_file( "alpha.txt", ios::in|ios::out ) ;
  if ( alpha_file.fail() )
  {
    cerr << "Unable to open alpha.txt" << endl ;
    return 1 ;
  }
  for ( c = 'A' ; c <= 'Z' ; c++ )
  {
    alpha_file << c ;
  }
  alpha_file.seekg( 11, ios::beg) ;
  alpha_file >> c ;
  alpha_file.seekp( 3, ios::cur ) ;
  alpha_file << c ;
  alpha_file.seekg( 0, ios::beg ) ;
  alpha_file >> c ;
  alpha_file.seekp( 0 ) ;
  alpha_file << c ;
  alpha_file.close() ;
  return 0 ;
}
```

7. A shift cypher is a simple method for encrypting text. The cypher works by replacing each letter in the text with the letter that occurs a certain distance from it in the alphabet. Non-alphabetic characters are not encrypted. For example, ABcD is encrypted to BCdE when the distance is 1; xYZ1 is encrypted to zAB1 when the distance is 2.

    Write a program to encrypt a text file with a distance specified by the user.

    An input file and an output file are required.

8. Write a program to merge two text files containing surnames into a single text file.

    Each line of the files contains a surname, followed by a space and a first name.

    Assume the files are sorted in ascending order of surnames.

9. Modify program P14M to allow for batch updates of the stock quantities in the file `stock.dat`. Each line of a text file `trans.txt` contains a stock code and a stock quantity. Read each line of `trans.txt` and update the appropriate stock record in `stock.dat`.

10. Overload the insertion operator << for the stock class in program P14M. The overloaded operator will display a stock record in the same format as lines 52 to 54 of P14M.

11. Modify program P14M to display a stock record, to delete a stock record and to add a new stock record. The program will require a menu:

```
1. Update Stock Quantity
2. Display a Stock Record
3. Delete a Stock Record
4. Add a Stock Record
5. Delete all Records
6. Display all Records
0. Quit

Enter option 0 to 6 :
```

Options 1 to 4 require the user to enter a stock code in the range 1 to 10.

The C++ code for options 1 and 6 already exist in the program.

Option 3 "deletes" a record from the file. A record is marked as being deleted by writing a 0 in the stock code field of the record in the file. A "deleted" record should not be displayed by options 2 or 6 or updated by option 1. Option 4 "adds" a new stock record to the file. To add a new stock record, use the stock code to calculate the file offset and read the record at this offset. If the stock code of the record from the file is 0, overwrite the stock file record with the new record. If the stock code of the stock file record is not 0, display an error message "stock code already exists".

Appendix A
List of C++ Keywords

Keywords are predefined reserved identifiers that have special meanings. They cannot be used as identifiers in a C++ program.

asm	auto	bad_cast	bad_typeid
bool	break	case	catch
char	class	const	const_cast
continue	default	delete	do
double	dynamic_cast	else	enum
except	explicit	extern	false
finally	float	for	friend
goto	if	inline	int
long	mutable	namespace	new
operator	private	protected	public
register	reinterpret_cast	return	short
signed	sizeof	static	static_cast
struct	switch	template	this
throw	true	try	type_info
typedef	typeid	typename	union
unsigned	using	virtual	void
volatile	while		

In addition to this list, consider all names beginning with an underscore (_) to be reserved for system use.

Appendix B
Precedence and Associativity of C++ Operators

Operator	Name	Associativity
::	Class scope	None
::	Global scope	
()	Function call	Left to Right
[]	Array subscript	
.	Member selection	
->	Member selection with a pointer	
++	Postfix increment	
--	Postfix decrement	
const_cast	Const cast	
dynamic_cast	Dynamic cast (checked at ruin-time)	
reinterpret_cast	Reinterpret cast (unchecked)	
static_cast	Static cast (checked at compile-time)	
typeid	Type identificatation (at run-time)	
-	Unary minus	Right to Left
+	Unary plus	
~	Bitwise NOT (ones complement)	
!	Logical NOT	
*	Indirection or dereference	
&	Address	
++	Prefix increment	
--	Prefix decrement	
new	Dynamic allocates memory allocation	
delete	Dynamic de-allocates memory	
sizeof	Size (in bytes)	
(type)	Cast (C-style)	
.*	Dereference	Left to Right
->*	Dereference	
*	Multiplication	Left to Right
/	Division	
%	Modulus	
+	Addition	Left to Right
-	Subtraction	
<<	Bitwise shift left	Left to Right
>>	Bitwise shift right	

(cont.)

Operator	Name	Associativity
<	Less than	Left to Right
<=	Less than or equals	
>	Greater than	
>=	Greater than or equals	
==	Equality	Left to Right
!=	Inequality	
&	Bitwise AND	Left to Right
^	Bitwise XOR	Left to Right
|	Bitwise OR	Left to Right
&&	Logical AND	Left to Right
||	Logical OR	Left to Right
=	Assignment	Right to Left
*=	Multiplication and assignment	
/=	Division and assignment	
%=	Modulus and assignment	
+=	Addition and assignment	
-=	Subtraction and assignment	
<<=	Bitwise shift left and assignment	
>>=	Bitwise shift right and assignment	
&=	Bitwise AND and assignment	
|=	Bitwise OR and assignment	
^=	Bitwise XOR and assignment	
? :	Conditional	Right to Left
throw	Throw an exception	Right to Left
,	Comma (sequence)	Left to Right

Notes:

1. Operators are listed in descending order of precedence. Where several operators appear in the same box, they have equal precedence.

2. Expressions within parentheses have a higher precedence than expressions without parentheses.

3. When an expression contains several operators with equal precedence, evaluation proceeds according to the associativity of the operator, either from right to left or from left to right.

Appendix C
ASCII Character Codes

A character code is a numerical value used to represent a character in the computer's memory. The ASCII (American Standard Code for Information Interchange) character set is defined as a table of seven-bit codes that represent control characters and printable characters. For example, the letter 'A' is represented by ASCII code 65, and '1' is represented by ASCII code 49.

The circumflex symbol ^ is used in the following table to indicate that the control key Ctrl is pressed simultaneously with another key.

Decimal code	Hexadecimal code		Key	Decimal code	Hexadecimal code	Key
0	00		^@	33	21	!
1	01		^A	34	22	"
2	02		^B	35	23	#
3	03		^C	36	24	$
4	04		^D	37	25	%
5	05		^E	38	26	&
6	06		^F	39	27	'
7	07	BEL	^G	40	28	(
8	08	BS	^H	41	29	)
9	09	TAB	^I	42	2A	*
10	0A	LF	^J	43	2B	+
11	0B	VT	^K	44	2C	,
12	0C	FF	^L	45	2D	–
13	0D	CR	^M	46	2E	.
14	0E		^N	47	2F	/
15	0F		^O	48	30	0
16	10		^P	49	31	1
17	11		^Q	50	32	2
18	12		^R	51	33	3
19	13			52	34	4
20	14		^T	53	35	5
21	15			54	36	6
22	16			55	37	7
23	17		^W	56	38	8
24	18		^X	57	39	9
25	19		^Y	58	3A	:
26	1A		^Z	59	3B	;
27	1B	ESC	^[	60	3C	<
28	1C		^\	61	3D	=
29	1D		^]	62	3E	>
30	1E		^^	63	3F	?
31	1F		^_	64	40	@
32	20		space			

(cont.)

Decimal code	Hexadecimal code	Key	Decimal code	Hexadecimal code	Key
65	41	A	97	61	a
66	42	B	98	62	b
67	43	C	99	63	c
68	44	D	100	64	d
69	45	E	101	65	e
70	46	F	102	66	f
71	47	G	103	67	g
72	48	H	104	68	h
73	49	I	105	69	i
74	4A	J	106	6A	j
75	4B	K	107	6B	k
76	4C	L	108	6C	l
77	4D	M	109	6D	m
78	4E	N	110	6E	n
79	4F	O	111	6F	o
80	50	P	112	70	p
81	51	Q	113	71	q
82	52	R	114	72	r
83	53	S	115	73	s
84	54	T	116	74	t
85	55	U	117	75	u
86	56	V	118	76	v
87	57	W	119	77	w
88	58	X	120	78	x
89	59	Y	121	79	y
90	5A	Z	122	7A	z
91	5B	[	123	7B	{
92	5C	\	124	7C	|
93	5D	]	125	7D	}
94	5E	^	126	7E	~
95	5F	—	127	7F	Del
96	60	'			

Appendix D
Fundamental C++ Built-in Data Types

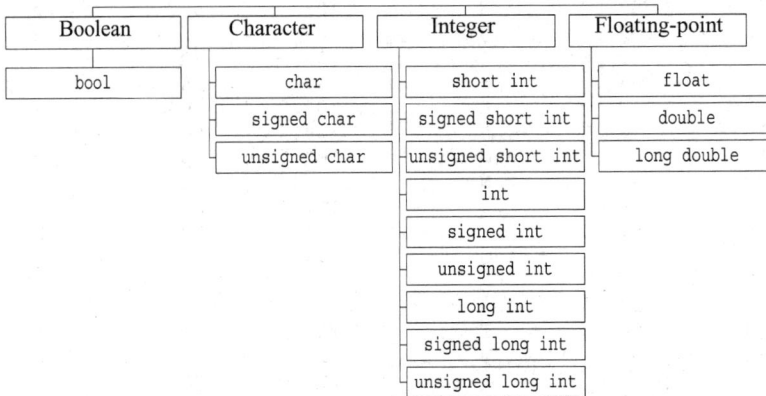

Boolean	Character	Integer	Floating-point
bool	char	short int	float
	signed char	signed short int	double
	unsigned char	unsigned short int	long double
		int	
		signed int	
		unsigned int	
		long int	
		signed long int	
		unsigned long int	

The range of values for each data type is compiler dependent.

Compile and run the C++ program data_types for details of the range of each data type on your system.

A sample run of the program is:

```
Data Type                         Bytes    Range
-----------------------------------------------------------------------------
bool                              1        0 to 1
char and signed char              1        -128 to 127
unsigned char                     1        0 to 255
short int and signed short int    2        -32768 to 32767
unsigned short int                2        0 to 65535
int and signed int                4        -2147483648 to 2147483647
unsigned int                      4        0 to 4294967295
long int and signed long int      8        -9223372036854775808 to 9223372036854775807
unsigned long int                 8        0 to 18446744073709551615
float                             4        1.17549e-38 to 3.40282e+38
double                            8        2.22507e-308 to 1.79769e+308
long double                       16       3.3621e-4932 to 1.18973e+4932
```

Appendix E
Common `iomanip` Manipulators

Manipulator	Input/Output	Description
endl	Output	Write a newline character and flush the output stream.
setw(w)	Input/Output	Set the field width to w. Effects the next item only.
setfill(c)	Output	Sets the fill character to c. The default fill character is a space.
setprecision(p)	Output	Sets the number of places of accuracy to p. (default value 6).
fixed	Output	Display floating-point values in decimal fixed-point notation.
scientific	Output	Display floating-point values in scientific notation.
skipws	Input	Ignore whitespace characters in input stream. This is the default setting.
noskipws	Input	Read whitespace characters in input stream.
showpos	Output	Precede positive numbers with +.
noshowpos	Output	Do not precede positive numbers with +.
flush	Output	Flushes the output stream. Forces all the data in the output stream to be physically written to the output device.
left	Output	Align left.
right	Output	Align right (the default).
showpoint	Output	A decimal point is displayed for a floating-point number.
noshowpoint	Output	A decimal point is displayed for a floating-point number only if the decimal portion is non-zero.
setbase(b)	Input/Output	Set the conversion base to b (8, 10 or 16).
showbase	Output	If the base is set to 8 (octal), a zero precedes the number. If the base is set to 16, 0x precedes the number.
noshowbase	Output	The number base is not shown.

`#include <iomanip>` is required for any manipulator that uses an argument.

`#include <iostream>` is sufficient for other manipulators.

Appendix F
Escape Sequences

Character	Meaning
\a	alert (bell)
\n	newline
\t	tab
\b	backspace
\r	carriage return
\f	form feed
\\	backslash
\"	double quotation mark
\'	single quotation mark
\0	null
\ddd	up to three-digit octal value
\xddd	up to three-digit hexadecimal value

Note: When an escape sequence appears in a string it counts as a single character.

Appendix G
The C++ Preprocessor

The preprocessor, as its name implies, processes a C++ program before it is read by the compiler. A C++ program is read by the preprocessor and modified by *preprocessor directives*.

C++ program ⟶ preprocessor ⟶ modified C++ program ⟶ C++ compiler.

Preprocessor statements are placed in a C++ program and tell the preprocessor how to modify the program. A preprocessor statement starts with a hash (#) and is followed by a preprocessor directive.

G.1 Including files

The `include` directive includes a specified file in the source file at the point at which the directive appears. The general format of #include is:

```
#include <file_name>
```

or

```
#include "file_name"
```

The angle brackets < and > in the first format instructs the preprocessor to search for the specified files in the standard include directory (folder) only. The standard include directory is where the standard include files such as `iostream` are stored.

The double quotation marks in the second format instruct the preprocessor to search for the file in the current directory and, if it is not found there, to search in the standard include directory. However, if a directory is specified within the double quotation marks then this directory only is searched.

By convention, header files written by a programmer have a file name with an extension of `.h`. System header files do not include a file name extension.

Here are some examples of `#include` directives:

```
#include <iostream> // Include iostream from standard include
                    // directory.
#include "my.h"     // Include my.h from the current directory
                    // and if not there look for the file in
                    // the standard include directory.
#include "\myinclude\my.h" // Windows uses a single backslash \.
#include "/myinclude/my.h" // UNIX and Linux uses a forward slash /.
                    // Include my.h from the directory
                    // myinclude. Only the specified
                    // directory myinclude is searched.
```

G.2 #define

Symbolic constants, such as the number of elements in an array (see line 9 of program P5B), have been defined in previous programs using `const`. It is also possible to define a symbolic constant using the

`#define` preprocessor directive. The format of `#define` is:

```
#define NAME replacement
```

where `NAME` is the symbolic constant. By convention, `NAME` is usually in uppercase characters. The preprocessor replaces all occurrences of `NAME` within the source file by `replacement` before the program is compiled. `NAME` must conform to the rules for constructing valid C++ identifiers.
Examples:

```
#define SIZE 10
#define DAYS_IN_WEEK 7
#define PI 3.141592653
#define END_OF_SENTENCE '.'
#define DIGITS "0123456789"
#define FIVE_SPACES "     "
#define END_OF_STRING '\0'
#define NEWLINE '\n'
#define BACKSPACE '\b'
```

Once a symbolic constant has been defined, it cannot be assigned to a different value without first removing the original definition. This can be done with the `#undef` directive. For example:

```
#define SIZE 10   // SIZE is 10.
#undef SIZE       // SIZE is no longer defined.
#define SIZE 20   // SIZE is now 20.
```

C++ programs generally use `const` rather than `#define` to define symbolic constants.

G.3　Conditional directives

The preprocessor conditional directives `#if`, `#else`, `#elif` and `#endif` can be used to include or omit blocks of C++ code. For example, suppose when debugging a program a programmer wants to check the values of three integer variables a, b and c at a certain point in the program. To do this, the following statements are placed in the program at the point where the variables are to be checked:

```
#if defined DEBUG
  cout << "a, b and c at this point are: " << a << b << c ;
#endif
```

If DEBUG has previously been defined (usually at the top of the program) with

```
#define DEBUG
```

the `cout` statement will be included in the program. If you leave out this `#define` or have DEBUG undefined with

```
#undef DEBUG
```

the `cout` statement will not be placed in the program. Debugging statements can be included or excluded in your program simply by defining or undefining DEBUG.
Symbolic constants can also be tested using the relational operators ==, !=, >, <, <= and >=.

```
#define DEBUG 'y'
...
```

```
#if DEBUG == 'y'
  cout << "a, b and c at this point are: " << a << b << c ;
#endif
```

The logical operators !, | | and && can also be used.

The conditional directives are commonly used to prevent multiple inclusions of a header file in a program. Class declarations are often defined in header files, and commonly contain preprocessor directives to prevent their multiple inclusions into a program.

In the following, a class called my_class is declared in a header file myclass.h. The file myclass.h typically contains the following:

```
#if !defined MY_CLASS_H
#define MY_CLASS_H

class my_class
{
  ...
} ;


#endif
```

The above preprocessor statements check to see if the symbolic constant MY_CLASS_H is defined. If it is, then the class has already been included in the program and is not included again. If not already defined, MY_CLASS_H is defined and the class is included in the program.